THE STEPHEN BECHTEL FUND

IMPRINT IN ECOLOGY AND THE ENVIRONMENT

The Stephen Bechtel Fund has

established this imprint to promote

understanding and conservation of

our natural environment.

The San Francisco Estuary Institute gratefully acknowledges the generous contributions to this book provided by

California State Coastal Conservancy

California State Water Resources Control Board

Friends of the Napa River

Napa County Flood Control and Water Conservation District

Napa County Watershed Information Center and Conservancy

Napa County Wildlife Conservation Commission

Napa Valley Vintners

The Napa Valley Historical Ecology Project is a collaborative effort of the San Francisco Estuary Institute and the Friends of the Napa River.

For more information about the San Francisco Estuary Institute, a research center providing region-wide science for ecosystem management, please visit www.sfei.org/HE.

For more information about the Friends of the Napa River, including educational programs and river events, please visit www.friendsofthenapariver.org.

NAPA VALLEY HISTORICAL ECOLOGY ATLAS

TABLE of AREA ETC					
Sq.M: Area	M:s Shore Line	M:s Inshore Line of Marshes	Miles of Creeks	Miles of Roads	Shore Line of Ponds
32,30	64,3T	38,3T	113,25	37,3T	3000

NAPA VALLEY

Exploring a Hidden Landscape

SAN FRANCISCO ESTUARY INSTITUTE

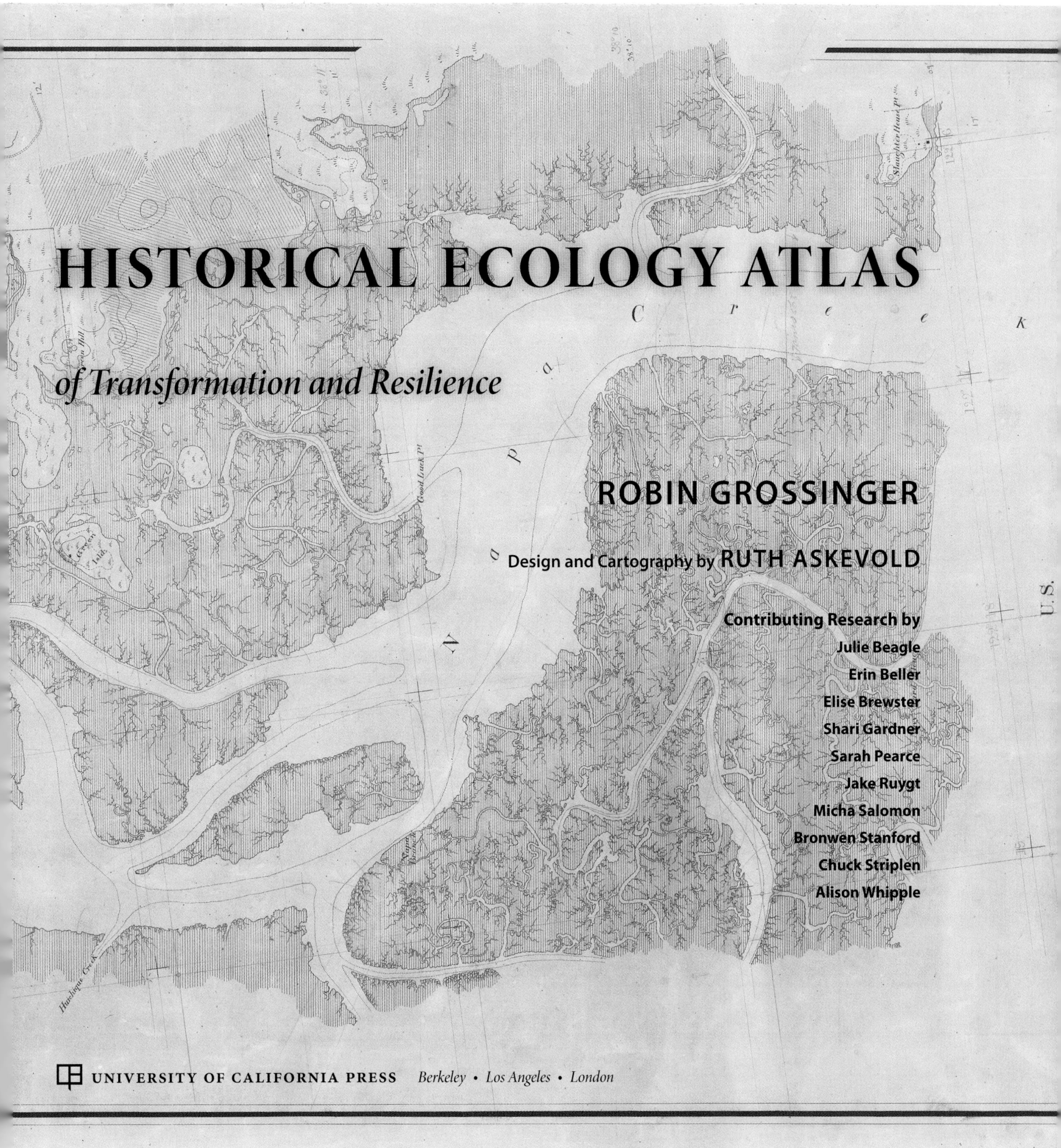

HISTORICAL ECOLOGY ATLAS

of Transformation and Resilience

ROBIN GROSSINGER

Design and Cartography by **RUTH ASKEVOLD**

Contributing Research by
Julie Beagle
Erin Beller
Elise Brewster
Shari Gardner
Sarah Pearce
Jake Ruygt
Micha Salomon
Bronwen Stanford
Chuck Striplen
Alison Whipple

UNIVERSITY OF CALIFORNIA PRESS   *Berkeley • Los Angeles • London*

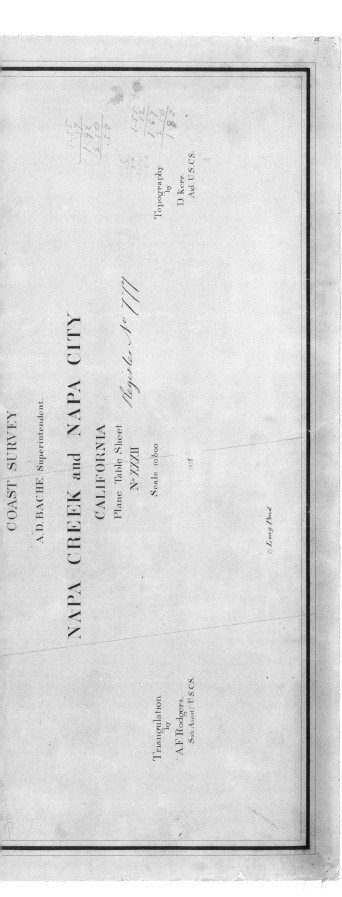

University of California Press, one of the most distinguished university presses in the United States, enriches lives around the world by advancing scholarship in the humanities, social sciences, and natural sciences. Its activities are supported by the UC Press Foundation and by philanthropic contributions from individuals and institutions. For more information, visit www.ucpress.edu.

University of California Press
Berkeley and Los Angeles, California
University of California Press, Ltd.
London, England

© 2012 by the San Francisco Estuary Institute

CONTRIBUTORS: *San Francisco Estuary Institute:* Julie Beagle, Erin Beller, Sarah Pearce, Micha Salomon, Bronwen Stanford, Chuck Striplen, and Alison Whipple. *Brewster Design Arts:* Elise Brewster. *Friends of the Napa River:* Shari Gardner. *Napa Botanical Survey Services:* Jake Ruygt.

CONTEMPORARY PHOTOGRAPHY: Ruth Askevold, Julie Beagle, Mark Defeo, David Hoffman, Jonathan Koehler, Herb Lingl, Lisa Micheli, Sarah Pearce, Jake Ruygt, Susan Schwartzenberg, and John White. All other contemporary photographs are by the author.

LIBRARY OF CONGRESS CATALOGING-IN-PUBLICATION DATA

Grossinger, Robin, 1969–.
    Napa Valley historical ecology atlas : exploring a hidden landscape of transformation and resilience / by Robin Grossinger ; design and cartography by Ruth Askevold ; contributing research by Julie Beagle ... [et al.].
       p. cm.
    Includes bibliographical references and index.
    ISBN 978-0-520-26910-1 (cloth : alk. paper)
    1. Landscape changes—California—Napa Valley—Maps. 2. Landscapes—California—Napa Valley—History—Maps. 3. Ecology—California—Napa Valley—History—Maps. 4. Natural history—California—Napa Valley—Maps. 5. Physical geography—California—Napa Valley—History—Maps. I. Askevold, Ruth. II. Title.

G1527.N3G4G7 2012
577.09794′190223—dc23

2011034718

Manufactured in Singapore

19 18 17 16 15 14 13 12
10 9 8 7 6 5 4 3 2 1

The paper used in this publication meets the minimum requirements of ANSI/NISO Z39.48–1992 (R 1997) (*Permanence of Paper*).

Title page illustration: The U.S. Coast Survey T-sheet of lower Napa River and its tidal marshlands, produced by surveyor David Kerr in 1858. *Kerr 1858, courtesy of NOAA.*

Cover illustration: (top) Napa Valley and River at Trancas Road, USDA 2005, courtesy of the USDA. (bottom) "Napa Valley and River" (detail), 1885, by Manuel Valencia, courtesy of the Collection of the Saint Mary's College Museum of Art, gift of James J. Coyle and William T. Martinelli.

To the people of Napa Valley, past, present, and future:

    The Napa, Caymus, Canijolmano, and Mayacma;

    David Kerr, for his inexplicably detailed map;

    today's stewards of its oaks, wetlands, and streams;

    and all those who will shape the valley in the years to come.

CONTENTS

	Acknowledgments	viii
	Preface	x
1	**EXPLORING THE NAPA VALLEY THROUGH TIME**	1
	Geography, Geology, Climate • A Brief History • A Map of the Napa Valley in the Early 1800s	
2	**OAK SAVANNAS AND WILDFLOWER FIELDS**	25
	Mapping the Oak Lands • Scattering Oaks • Burning the Valley • Savanna Composition • A World of Wildflowers • Valley Oaks and People • Oak Trajectories • Re-Oaking the Valley	
3	**CREEKS**	49
	Spreading Streams • The High Water Table Problem • Plumbing the Valley • Riparian Loss and Recovery • Undercut Trees and Channel Incision • Braided Channels • Summer Water	
4	**VALLEY WETLANDS**	67
	Valley Freshwater Marsh • Wet Meadows • Vernal Pool Complexes • Alkali Meadows • Wetland Fragmentation and Resilience	
5	**NAPA RIVER**	81
	The River Spread into Wetlands • Islands and Sloughs • The Valley in Flood • Has the River Been Straightened? • How Deep Was the River? • Disappearing Gravel Bars • Did the River Flow Year-Round? • Riparian Forest Width • Narrowing of the Riparian Corridor • Beavers and Snags • Jepson's Napa River • Glimpses of the Tidal River Corridor • Fish Assemblages • Patterns of River and Valley • The River Responds	

6 TIDAL MARSHLANDS 121

Kerr's Map • The Tule Lands • Dredging the Tidal River • Reclamation and Restoration • Marsh Chronologies • The Edge

7 LANDSCAPE TRANSFORMATION AND RESILIENCE 143

Using Historical Ecology • Resilient Landscapes • Redesigning the Valley

8 LANDSCAPE TOURS 153

South of Napa: Vallejo, American Canyon, Carneros • Lower Valley: Napa, Oak Knoll, Yountville • Mid-Valley: Oakville, Rutherford, St. Helena • Upper Valley: Larkmead, Calistoga

Common and Scientific Names of Species   179

Notes   181

Bibliography   197

Index   215

ACKNOWLEDGMENTS

We would like to recognize Josh Collins, David Graves, Chris Malan, and Luna Leopold, whose vision initiated the Napa Valley Historical Ecology Project. The Atlas would not have come to fruition without the support and guidance provided by Leigh Sharp of the Napa County Resource Conservation District (NCRCD); Patrick Lowe, Jeff Sharp, and Hillary Gitelman of the Napa County Conservation, Development, and Planning Department; Bernhard Krevet, Francie Winnen, and the Board of Friends of the Napa River (FONR); Jeremy Sarrow and Rick Thomasser of the Napa County Flood Control and Water Conservation District; Mike Napolitano of the San Francisco Bay Region Water Quality Control Board; Amy Hutzel and Betsy Wilson of the California State Coastal Conservancy; Phill Blake of the Natural Resources Conservation Service; Rex Stults of Napa Valley Vintners; Rainer Hoenicke and Lester McKee of the San Francisco Estuary Institute; Mike Connor; Herb Schmidt; Kent Lightfoot; and Les Rowntree.

Many people have generously contributed information and expertise to this effort, including Bill Lyman; Sharon Cisco Graham; Garrett Buckland; Stephen Rae; Al Edmister; Tom Wilson; Davie Piña; Tom and Olga Shifflett; Jonathan Koehler, Bob Zlomke, and Mike Champion of the NCRCD; Babe and Sandra Learned; Bill Grummer; Jane Slatterly and Ginny Leija of the Napa County Department of Public Works; Ralph Ingols, Judith Sears, and Jim Hench of FONR; Jeri Gill; Isabelle Minn and Melissa Erikson of Design, Community, and Environment; Eric Gerhart; Clayton Creager; T. Beller; Ann Baker; Steven Hasty; John Cloud; Kris Tacsik; David Garden; Carl Larson; Ira and Shirley Lee; Al and Mavis Fournier; Betty Mukerji; and Lee Hudson. We benefited from the able assistance of Kristen Cayce and Marcus Klatt of SFEI, and Maika Nicholson, who helped bring the Atlas to completion through a summer internship from The Bill Lane Center for the American West at Stanford University.

We are also indebted to a number of colleagues for sharing their insights and expertise: Robert Leidy for his insights into historical river characteristics; Jonathan Koehler for discussions about fish assemblages and the use of his table in Chapter 5; Todd Keeler-Wolf for help interpreting sycamore and oak distributions; Jennifer Natali for her inspired reconstructions of river morphology illustrating Chapter 5; Stephen Rae for coring oaks, no easy task; Mike Napolitano and Andy Collison for stimulating discussions about river morphology; and Susan Schwartzenberg for her incisive observations and photographic perspective on field expeditions. The Atlas was improved by review comments from Peter Baye,

Brian Cluer, Frank Davis, Gretchen Hays, Rainer Hoenicke, Jonathan Koehler, Bernhard Krevet, Robert Leidy, Amber Manfree, Laurel Marcus, Mike Napolitano, Mike Rippey, Leigh Sharp, David Steiner, and Lin Weber. The project benefited from the strong scientific community engaged in understanding the Napa River and its watershed.

We would also like to thank the talented team at UC Press, including Reed Malcolm, for helping develop the initial idea, and Blake Edgar, Kate Hoffman, Hannah Love, Lynn Meinhardt, and Claudia Smelser for helping turn the idea into reality. Finally, Robin would like to thank the extended members of the weekend research team, Erica, Leo, and Joey.

PREFACE

As landscapes change, people forget what was once there. In the Napa Valley, it is still less than 200 years since native peoples managed the land, coexisting with populations of tule elk, grizzly bears, beavers, and salmon. Yet little knowledge about that historical landscape persists. Memories have disappeared, and nobody really knows how the valley looked just three lifetimes ago. Similarly, the history of modifications made by generations of hard-working newcomers to the valley is hazy. Long forgotten, many of these changes now seem natural or permanent elements of the landscape, rather than recent alterations.

Despite the dramatic transformation of the past two centuries—and the Napa Valley has changed more than might be expected for an agricultural region—the links to the past have not been completely severed. From a historical perspective, the changes are still fresh. Many aspects of the historical landscape are still visible if you look closely.

When rediscovered, the forgotten habitats and functions of the historical Napa Valley can help us understand current environmental challenges and identify practical options for the future. Napa Valley towns were built in the shade of celebrated valley oak groves. Might these elegant, drought-tolerant trees make sense as shade trees once again, as we anticipate the possibility of hotter summers and scarcer water? As we try to resolve persistent flooding and bank erosion, might we want to consider the ways the historical river modulated high flows? What role might beavers, salmon, and wood ducks play in the 21st century? Before we can apply lessons from the past to our modern landscape, we must first get to know the historical landscape. Historical ecology provides the tools to reconstruct that world.

The Napa Valley is a strikingly beautiful place. It was considered exceptionally so in the 1850s as well, albeit for largely different reasons. It is a region with a strong sense of place and a deep appreciation of the land. Yet many of the landscape characteristics that so impressed early visitors are nearly gone, their absence unnoticed. Without special efforts, the natural heritage of the valley will continue to decline.

Landscapes are always changing. What will the Napa Valley of the future be like? What patterns of tree and field, wetland and river, road and town will be most resilient to changing climates? What landscape will be most appealing to inhabit and visit? How do we ensure that the beauty and bounty of the 19th and 20th centuries continue through the 21st?

The next pages reconstruct these hidden dimensions of the land. Inevitably, this is a story of extreme change. Yet the historical landscape is not as far away as it might seem. As it comes into focus, we see that not all is gone and much could come back if we wanted it to. With strategic efforts, many elements of the recent past could be reintegrated within the existing cultural and economic framework of the valley. Remarkably, the historical landscape helps us recognize the persistence and resilience of the living landscape we inhabit today.

Images of the Napa River and city of Napa. Successive data sets show both change and continuity. Produced for different purposes with different techniques, each view provides a body of evidence about the nature of Napa Valley. *From left to right: Kerr 1858, courtesy of the National Oceanic and Atmospheric Administration (NOAA); U.S. Department of Agriculture (USDA) 1942, courtesy of the Napa County Resource Conservation District (NCRCD) and the Natural Resources Conservation Service (NRCS); USDA 2009, courtesy of the USDA; Tracy 1858d, courtesy of The Bancroft Library, UC Berkeley.*

1 • EXPLORING THE NAPA VALLEY THROUGH TIME

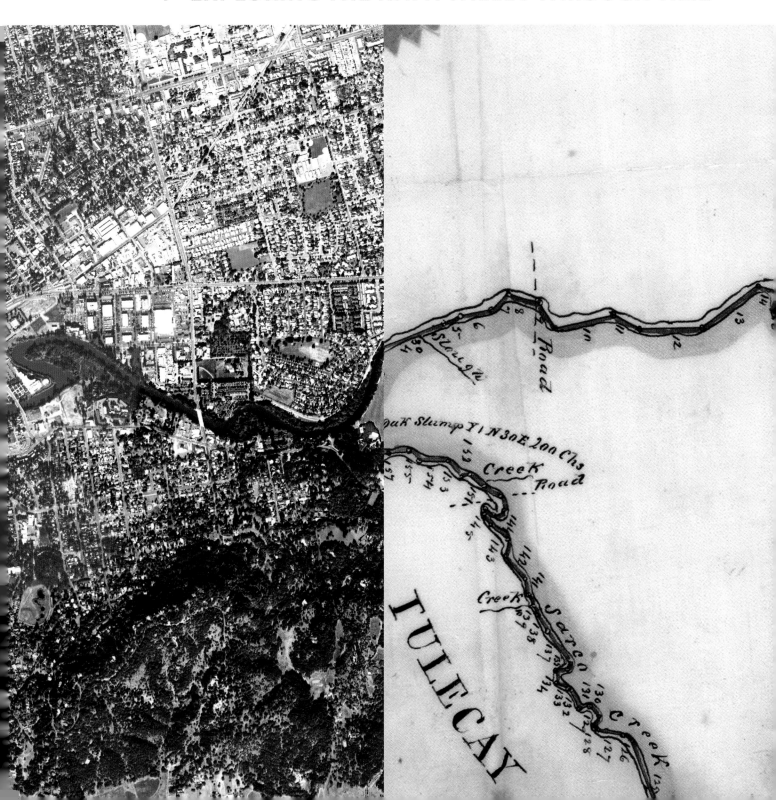

Contemporary landscapes present a puzzle. Because of the intensity of recent land use, underlying patterns and processes are difficult to interpret, overwhelmed by a maze of buildings and roads, culverts and levees. As we try to improve the health of our rivers, wetlands, and woodlands we often don't know how they used to work and how they have changed. Our standard scientific tools, powerful as they may be, are relatively ineffective at unraveling the past few centuries of landscape transformation.[1]

Fortunately, a diverse cast of surveyors, writers, artists, and photographers left records of the landscapes they saw. Looking through their eyes, we can begin to fill the gap in our collective memory, to reconstruct the landscapes of the not-so-distant (yet mostly forgotten) past.

Historical ecology synthesizes this jumble of information into a usable format, integrating the clues provided by archival data with persistent physical evidence such as soils, remnant trees and wetlands, and topographic patterns. Given the heterogeneous and often incomplete nature of these data, historical ecology can be a risky enterprise, fraught with the potential for misinterpretation, misdirection, and nostalgia.[2] But when successful, historical ecology can transform our understanding of the familiar places we live, work, and play. Through the examination and inter-calibration of hundreds of independent documents, the history of these landscapes emerges from the shadows.

Historical ecology—literally the study of past ecology—emerges from a long line of investigations within disciplines as varied as botany, ecology, geology, geography, anthropology, and history.[3] In 1864, George Perkins Marsh's *Man and Nature*, a formative text in the field of ecology, aimed "to indicate the character and, approximately, the extent of the changes produced by human action in the physical conditions of the globe we inhabit." In the 1920s, geographer Carl Sauer laid out the approach to discovering the "facts of the natural landscape" and "the amount and character of transformations induced by culture" through the field of historical geography.[4] Sauer even created a guide to enable children to explore the historical ecology of their neighborhoods, including finding old maps, identifying old trees, and interviewing longtime residents.[5] But while the accurate recognition of human-induced landscape change underlies the science and ethic of ecology and environmental conservation, historical research tools have often been overlooked by environmental researchers.[6]

In recent decades, however, efforts to restore landscape functions at a regional scale have driven an expansion in historical ecology methods and

Views of a changing landscape. The distinctive meander on the Napa River known as the "Oxbow" (top) in 1858 and 2003. Orchards were replaced by vineyards while streamside vegetation expanded at Oak Knoll (middle) between 1908 and 2006. Oaks in a field near Calistoga (bottom) in 1942 and 2010. *Top left: Kerr 1858, courtesy of NOAA. Top right: photograph courtesy of Mark Defeo and the Napa County Flood Control and Water Conservation District. Middle left: Darms 1908. Middle right: photograph January 24, 2007. Bottom left: USDA 1942, courtesy of the NCRCD and NRCS. Bottom right: USDA 2010, courtesy of the USDA.*

applications. Environmental historians have investigated how people have interacted with and shaped their landscapes. Fluvial geomorphologists have analyzed the physical transformation of rivers and floodplains. Ecologists and geographers have reconstructed vegetation change in landscapes as varied as the Wisconsin forests, the Southwest plains, the English countryside, and Manhattan.[7]

In California, as in other parts of the world, historical ecology has become an important tool for increasing our fundamental understanding of ecosystems. Historical ecology studies at the San Francisco Estuary Institute (SFEI) began in the early 1990s to address the need for reliable historical landscape analyses to guide restoration strategies for San Francisco Bay wetlands.[8] Since that time, SFEI and partners have initiated an array of historical ecological investigations throughout California. Abundant local interest in the nature of the past landscape led to the initiation of the Napa Valley Historical Ecology Project in 1999, in partnership with Friends of the Napa River.

ATLAS STRUCTURE

This first chapter introduces the study of historical landscapes. After discussing the origins and purpose of historical ecology, we review the basic geologic and climatic setting. We next take a tour through recent human history, including the chroniclers whose work we use to reconstruct the past landscape. After these introductions to the landscape and its history, we begin our virtual journey to the historical Napa Valley, starting with a composite map reconstructing the valley as it looked two centuries ago.

Chapters 2 through 6 investigate the distinct stories of the five major components of the valley landscape in the early 1800s, focusing on how these systems are revealed through historical documents, how they have been used and altered, what remains, and what elements might be relevant in the future. We start with the highest and driest parts of the valley: the oak savannas, grasslands, and wildflower fields (Chapter 2). We then follow the creeks (Chapter 3) downslope into the valley wetlands (Chapter 4). Napa River (Chapter 5) unites these features and flows downstream to the tidal marshlands (Chapter 6) at the foot of the valley. The landscape chapters are structured around two-page spreads, each focused on a particular topic.

Chapter 7 considers overall changes to the valley and future scenarios. In Chapter 8, we explore the valley through a set of tours of contemporary remnants. The tours run from south to north, visiting forty-two sites that illustrate the historical landscape and its expression in the valley today.

GEOGRAPHY, GEOLOGY, CLIMATE

The Napa Valley is one of coastal California's prominent tectonically controlled, northwest-trending valleys (others include the Salinas, Santa Clara, and Sonoma valleys). Its underlying form originated during relatively recent geologic history—about three million years ago—from the collision of tectonic plates. As the Pacific and North American plates produced the surrounding mountains, the area between was forced downward (a downwarp or syncline) to create the elongate Napa Valley.[9]

The resulting long and narrow shape—less than 4 miles wide for most of its length—makes the Napa Valley a particularly well-defined place: north of St. Helena, the adjacent ranges often converge within a mile. But at its lower, southern end, the valley widens to San Pablo Bay, connecting to the broader San Francisco Bay Area. The Napa Valley lies only 50 miles from San Francisco by land (and 45 by water)—a position sufficiently distant from the Bay Area's urban center to avoid early urbanization, but close enough to serve as a food source and vacation destination since the mid-19th century.

The valley's relatively flat, contained geography masks substantial variation. In less than 30 miles, the valley slopes 345 feet from the resort town of Calistoga (365 feet elevation) to the city of Napa (20 feet), positioned for tidal access. (By way of contrast, the Sacramento Valley requires 140 miles to gain less than 300 feet between Sacramento and Red Bluff.) Topographic variation within the valley explains many of its patterns of ecology, agriculture, and human settlement.

The Napa Valley's Mediterranean climate has been critical to its success as a premier wine-growing region. Substantial winter rainfall, warm summer days, and cool nights provide the basic conditions for successful grape cultivation, even without irrigation.[10] The valley experiences winter rains and an annual summer drought: little rain falls between May and September. Within this reliable seasonal framework, rainfall amounts and timing vary dramatically from year to year (over 20 floods and several major droughts since 1850) and geographically. Average annual rainfall varies from 25 inches in the city of Napa to 35 inches in St. Helena, with much higher values reported in the western hills. Interestingly, this substantial range of precipitation does not appear to have directly controlled the distribution of major habitats, which occurred broadly along the rainfall gradient.

EXPLORING THE NAPA VALLEY • 7

Napa Valley places. (opposite page) The map shows the Napa Valley (in orange) in relation to major roads, creeks, and cities. The boundary of the Napa River watershed is shown in dark gray. Tribal areas are taken from Milliken 2009 and shown in shaded boxes. Maps on this page and throughout the Atlas are oriented with north at top of map unless otherwise noted.

Napa Valley in the San Francisco Bay Area. (top) The Napa Valley study area, shown in orange, includes the alluvial and tidal lands along the Napa River. *Satellite view courtesy of Esri.*

Geology of the Napa River watershed. (bottom) Volcanic materials dominate the upper watershed, with smaller constituents of the more erosive Franciscan Complex and Great Valley Sequence. Younger alluvial deposits form the valley floor. Bay muds and marsh deposits underlie the tidal lands that extend the valley toward the Bay. *Data from Graymer et al. 2006; classification after Sloan 2006.*

Artificial Fill

Quaternary Alluvium

Quaternary Hillslope Deposits

Quaternary Bay Mud and Marsh

Quaternary/Tertiary Sedimentary Rocks

Tertiary Sedimentary Rocks

Tertiary Volcanic Rocks

Franciscan Complex

Great Valley Sequence

Temperature also varies through the valley. Cooler temperatures are found in the fog-influenced southern end, while warmer conditions prevail in the mid- and upper valley. The valley's intermediate position between the cooling coast and hotter Central Valley makes for conditions particularly conducive to grape cultivation. (This position may actually minimize warming in response to global climate change, as hotter inland conditions draw more cooling fog into the Napa Valley.[11]) The well-watered, sloping Napa Valley contributes a significant amount of freshwater to San Francisco Bay, ranking as the Bay's second largest tributary downstream of the Sacramento-San Joaquin Delta.

While tectonic activity created the space for the Napa Valley, its physical surface is the incremental product of seasonal movements of sediment and water. During winter rains, upland streams gather rocks, sand, and silt from the mountains and carry these materials through the canyons to the valley. In flood, the streams spread out and migrate across the valley, depositing sediment in elaborate patterns to build the valley floor. Over thousands of years, the streams have constructed a rippling surface of alluvial (stream-built) topography hundreds of feet thick, gently sloping from the canyon mouths to the bottomlands of the valley, superimposed on the valley's downward slope to the Bay.

At its lower end, the Napa Valley meets the brackish waters of San Francisco Bay, where tidal dynamics direct the deposition of sediment to form marshlands. Here the valley is gradually being submerged by the rising seas, which have been naturally expanding since the last Ice Age. Stretching between the uplifting, eroding mountains and the rising Bay, the valley exists in a dynamic state, continually reshaped by the fluxes of land and water at either end.

In some ways this landscape of widely flooding streams, unpredictable rainfall, and rising tides is quite unstable. Yet these dynamic processes have also created fertile land, reliable groundwater, and an abundance and diversity of habitats. Plants, animals, and human cultures have found ways to live in the valley in patterns that have persisted for hundreds, and sometimes thousands, of years. Not all of this change happens in the same place at the same time. In fact, the landscape is highly structured, focusing the forces of water and sediment into distinctive patterns that are reliable for long periods of time. The locations of trees, streams, and wetlands are guided by these patterns. Human activity, when most sustainable, is well calibrated to this landscape framework. Studying the shapes of hills, fans, rivers, and wetlands reveals the ecological fundamentals of a place, giving us clues about how we can live there most sustainably.

Mountains and Knolls

Rough, geologically young mountain ranges surround the Napa Valley in all directions but the south.[12] Their wrinkled shapes determine local wind patterns and rainfall, defining watersheds and giving rise to the individual creeks that enter the valley. The bedrock of the mountains provides the raw parent material that washes downstream to form the productive valley floor.

Unlike most of the San Francisco Bay Area, these ranges are primarily of volcanic origin, produced by volcanoes active several million years ago that deposited materials on top of sedimentary formations.[13] (Calistoga's hot springs, the Palisades rock formation, and the obsidian deposits of Glass Mountain are all remnants of this activity.) While the valley's watershed is dominated by relatively hard volcanics, there are several large areas of the sedimentary Franciscan formation in the middle of the valley. These more erosive rocks produce the valley's two largest braided stream channels, Conn and Sulphur creeks. This local geologic variation may also contribute to the character of the valley's great Oakville and Rutherford wines.[14]

An unusual characteristic of the Napa Valley is its scattered knolls—small, isolated hills protruding from the valley floor. Geologists have hypothesized that these bedrock islands may be the product of ancient megaslides.[15] In places, the knolls redirect the course of the Napa River or constrict drainage to form wetlands, locally altering the character of the valley.

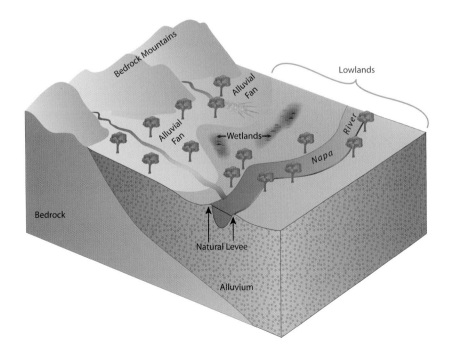

Conceptual model of Napa Valley landforms and surface drainage. Streams create alluvial fans radiating broadly from their canyon mouths. Because of the localized sediment input, these lands build up slightly higher than the lowlands. The lowlands receive overflow from both adjacent creeks and the Napa River. The natural levees along the river, which block surface drainage, and the dissipation of many creek channels in their lower reaches both contributed to the seasonal spreading of water across the lowlands.

Fans and Lowlands

A dominant element shaping the valley floor is the pattern of alluvial fans—gently sloping cones of sediment radiating outward from canyon mouths, created as the streams swing back and forth across the valley. Watersheds of different size, geology, and runoff create differently sized fans. Adjacent watersheds create converging fans, often with a lower area of trapped drainage in between, where wetlands form.[16]

This undulating surface—with crests a few feet higher than neighboring lows—is easily overlooked. Yet these critical differences largely determine the quality of land for farming, the character of grapes, the risk of flooding, and the best places to live. More than rainfall and temperature, the fans determine where trees grow and wetlands form. They push the river around and form areas with different drainage characteristics. As we will see, they explain much of the history and nature of the valley.

The interplay between fans and lowlands shapes much of the diversity of the valley floor. Alluvial fans provide slightly higher, well-drained ground—the famous grape-growing "benches"—before dissipating into the floodplains and lowlands along the Napa River. Towns and roads were mostly built on the fans, safely above major floods. The relatively porous fans percolate water into the groundwater basin—water which reemerges in the lowlands to support perennial streams and wetlands. In contrast, the lowlands are more poorly drained and prone to flooding.[17] Broad riparian (streamside) forests and widely branching side channels were found where the lowlands widen in the absence of large fans.

Coalescing alluvial fans are a distinctive feature of the relief in all the lowland areas.
—E. J. CARPENTER AND STANLEY COSBY, 1938

Looking upvalley toward Mt. St. Helena. This view looks northwest across the Oakville and Rutherford districts from about 2,000 feet altitude, showing the Napa River, the valley, and the adjacent, converging mountain ranges. The spring scene (March 2007) includes trees without leaf along the river and yellow patches of wild mustard in vineyards; Highway 29 is on the left.
©Herb Lingl/aerialarchives.com

A BRIEF HISTORY

While the Napa Valley has existed for several million years, human presence in the region is comparatively recent. Yet when the Portolá expedition entered the San Francisco Bay Area in the fall of 1769, native Californians had lived in the Napa Valley for several thousand years. Large villages were located in the vicinity of today's major settlements: the cities of Napa, Yountville, St. Helena, and Calistoga. Anthropologists and historians have estimated that three to four thousand people lived in the valley at the time of European contact, in four distinct tribes or bands: the Napa, the Caymus, the Canijolmano, and the Mayacma.[18] At least two language groups, Patwin in the south and Wappo to the north, were represented in the valley.[19] The valley was not a wilderness but rather a landscape inhabited by long-standing cultures.

Native peoples, it has been increasingly recognized, did not live passively on the land but actively shaped its nature.[20] The Napa, Caymus, Canijolmano, and Mayacma gathered grass and wildflower seeds, collected acorns and stored them in granaries, and shaped plant communities through pruning, transplanting, and burning. In fact, the local tribes were probably responsible for one of the most fundamental characteristics of the historical Napa Valley: its openness. By purposefully burning the meadows and savannas, native peoples helped create the inviting landscape that enabled Spanish, Mexican, and American ranching culture, as well as the quick transformation to American agriculture. Having worked the land for generations, native peoples had a detailed knowledge of landscape characteristics and dynamics—information that was largely lost as local culture was decimated after European contact.[21]

In 1776, the first Spanish mission in the Bay Area was established in San Francisco, less than 25 miles from the mouth of the Napa River. Yet for several decades Spanish influence on the northern margins of San Francisco Bay was hindered by the natural deepwater barriers of the Golden Gate and Carquinez Strait. As late as 1810—by which time tribes in the southern and central parts of the Bay Area had been heavily impacted—native cultures of the North Bay remained largely intact.[22] In the next decade, however, mission establishment efforts focused on the north. It is during this decade, just two centuries ago, that colonization began to disrupt the indigenous Napa Valley.

By 1815, the lower Napa Valley probably experienced significant population decline due to mission recruitment; impacts continued northward through the valley during the 1820s.[23] Despite the devastating

Land-use timelines. (following pages) The Napa Valley has experienced a succession of rapid land-use changes during the past two centuries. The population of the valley has increased greatly, particularly in recent decades, but population trends vary by city. For example, while American Canyon has doubled in population during the past decade, the censuses of St. Helena and Yountville, each located within the agricultural preserve, have decreased slightly during the same time. To compare relative growth among cities, the graph shows the percent of each city's population relative to the 2010 population. Actual 2010 population values are in parentheses.[24]

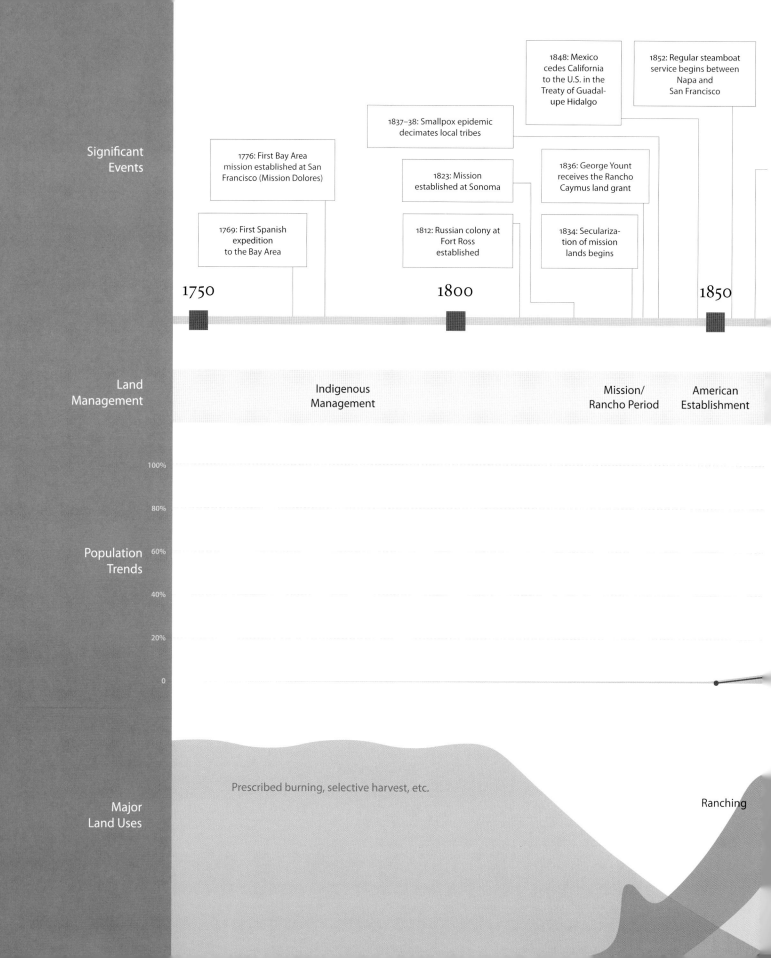

EXPLORING THE NAPA VALLEY • 13

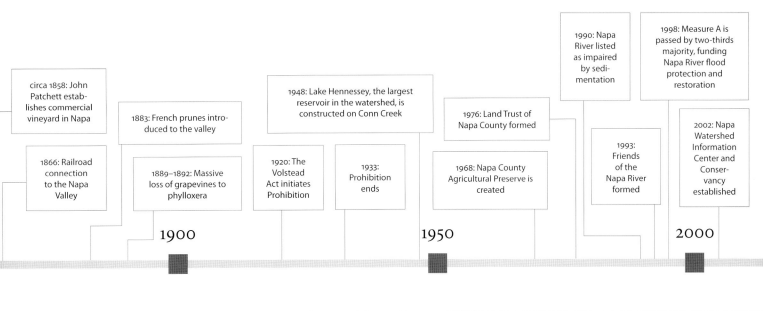

- circa 1858: John Patchett establishes commercial vineyard in Napa
- 1866: Railroad connection to the Napa Valley
- 1883: French prunes introduced to the valley
- 1889–1892: Massive loss of grapevines to phylloxera
- 1920: The Volstead Act initiates Prohibition
- 1933: Prohibition ends
- 1948: Lake Hennessey, the largest reservoir in the watershed, is constructed on Conn Creek
- 1968: Napa County Agricultural Preserve is created
- 1976: Land Trust of Napa County formed
- 1990: Napa River listed as impaired by sedimentation
- 1993: Friends of the Napa River formed
- 1998: Measure A is passed by two-thirds majority, funding Napa River flood protection and restoration
- 2002: Napa Watershed Information Center and Conservancy established

Agricultural Intensification and Transitions

Modern Agriculture and Urban Development

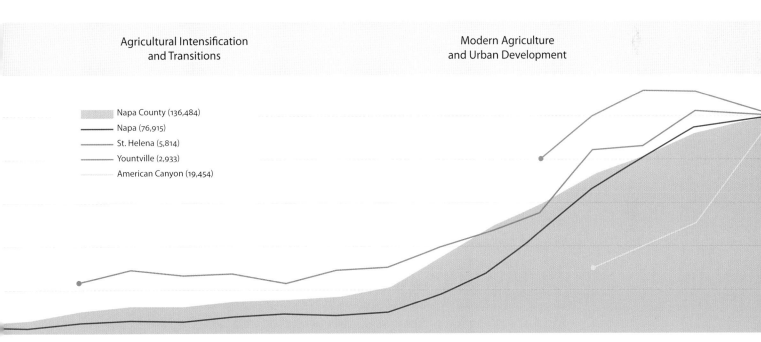

- Napa County (136,484)
- Napa (76,915)
- St. Helena (5,814)
- Yountville (2,933)
- American Canyon (19,454)

Grain

Vineyards

Orchards

Urban/Suburban

effects of land dispossession and epidemic disease, some traditional practices such as controlled burns continued for decades.[25] A reminder of this prolonged period of cultural contact is the contemporary use of native place names in the North Bay. The indigenous Napa, Sonoma, Petaluma, Carquinez (and, in the Napa Valley, Tulucay, Soscol, and Caymus) stand in contrast to the Catholic saints' names that predominate in regions of California with more-rapid Spanish colonization.

Russian trappers and traders also likely visited the Napa Valley from their colony at Fort Ross, established in 1812. But the valley's colonial experience was notably brief, as it lay at the edge of both the Spanish and Russian expansions. The valley served for several years as a secondary ranch for the mission at Sonoma, established in 1823, but the mission system disintegrated in 1834. By 1840 the area had been redistributed by the Mexican government to several favored Mexican citizens and American expatriates, introducing lasting names such as Yount, Bale, Higuera, and Vallejo. Thousands of cattle spread across the valley as part of the Mexican ranchos, and the area saw limited farming and water diversions near the rancho homesteads. The rancho era, too, was relatively short-lived in comparison with lands to the south.

Following the Mexican-American war (1846–48), cattle ranching expanded and then was largely replaced by farming. In 1852, traveler John Russell Bartlett observed the rapid shift: the valley "was fast being brought into cultivation…every farmer was occupied in marking out his land…[p]loughs were cutting up the virgin sward in all directions."[26]

We left…by the way of the Napa Valley with its meadows, orchards, and vineyards.
—SOIL SCIENTIST MACY LAPHAM DESCRIBING THE VALLEY IN 1936[30]

American settlers began to reshape the valley, planting vast grain fields and plowing the land to access soil moisture. New channels were constructed to drain wetlands and contain creeks to protect fields and towns. As early as 1870, a few levees were constructed along the Napa River to reduce flooding. By the 1880s, wheat and other grains began to give way to higher-value vineyards and orchards, to the detriment of the oak savannas.[27] Six thousand Napa Valley prune trees in 1880 expanded to more than one million by 1930, while wheat cultivation in the county decreased from 33,000 acres to just 3,000 acres during the same period.[28] Vineyard extent declined in the 1890s because of the rootstock disease phylloxera, but recovered fairly quickly.[29] The turn of the 20th century thus saw an intensively farmed, mixed agricultural landscape: prune orchards and vineyards with scattered pear orchards, walnut orchards, and dairies. The tidal marshlands south of the town of Napa were being diked off for agriculture. Pressure on the Napa River's tidal reaches for commercial transportation led to the establishment of an Army Corps dredging program, greatly altering the river's form and function. The expansion of the stream network for drainage and flood protection continued. Yet crops were dry-farmed; the moist valley soils in fact supported a successful agricultural industry without irrigation for over a century, into the 1960s. (In contrast, groundwater levels in the Santa Clara Valley fell by over 100 feet by the 1960s as a result of 20th-century agricultural irrigation.[31])

In 1940, the combined population of Napa Valley towns was still just 26,000 people—much closer to pre-European contact levels than to today's 136,484.[32] The rate of residential expansion steepened dramatically around World War II—first to accommodate workers from Mare Island Naval Shipyard in Vallejo—and has continued steadily since. To supply water to this growing population, the largest reservoir in the watershed (Lake Hennessey) was constructed during the late 1940s.

In 1968, recognizing the surrounding pressures for urban and suburban development, voters enacted the unprecedented Napa County Agricultural Preserve, which has maintained the agricultural character of the valley. As a result, the rate of population growth in Napa Valley has been much slower

Bird's-eye view of Napa City, looking west. A 19th-century depiction of the growing city of Napa, built on the banks of the Napa River. *Courtesy of The Bancroft Library, UC Berkeley.*

A TALE OF TWO VALLEYS: NAPA AND SANTA CLARA

Two centuries ago, Napa and Santa Clara valleys looked quite similar. Early explorers commented on the graceful oak savannas and the vast marshlands at the Bay margin.[36] In the first aerial photographs, taken around 1940, the two valleys still appear almost identical. Orchards cover fertile soils, interspersed with vineyards, pastures, and homesteads. Santa Clara Valley—the Garden of Heart's Delight—was perhaps the more famous agricultural area. Just three decades later, their trajectories had veered apart: Santa Clara Valley turned into Silicon Valley; Napa Valley became *Napa*.

The presence of Stanford University, Moffett Field, and nearby population centers catalyzed the transformation of Santa Clara Valley into the global epicenter of high-tech industry, while Napa became a renowned wine region. (Leland Stanford considered building his school in Calistoga; would Napa have become Silicon Valley?) The establishment of the Agricultural Preserve in 1968, the first in the country, reflected the local desire to preserve the agricultural landscape. Without the presence of a high-value agricultural economy, and ongoing efforts to preserve it, Napa would likely have followed a similar trajectory to Santa Clara Valley and most other agricultural areas around the Bay.

Four square miles, Napa and Santa Clara. Photographs of Napa Valley (east of Yountville, top) and Santa Clara Valley (southeast San Jose, bottom) show similar landscapes of orchard and pasture around 1940. Corresponding views less than seven decades later show conversion to vineyards (Napa) and neighborhoods (Santa Clara)—each within the same framework of 19th-century roads and property lines. Aerial imagery throughout the Atlas includes 2005, 2009, and 2010 data from the National Agricultural Imagery Program (NAIP) and a photomosaic created from over 80 early aerial photographs from the collection of the Napa County Resource Conservation District (NCRCD) and Natural Resources Conservation Service (NRCS).[37]

than that in most of the rest of the Bay Area. Nevertheless, population and urban areas have still increased substantially in recent decades, particularly outside of the agricultural preserve. While Calistoga, St. Helena, and Yountville each decreased in population slightly between 2000 and 2010, American Canyon doubled in size (from 9,774 to 19,454).

The most recent wave of viticultural success began in the 1960s, allowing the industry finally to recover from the effects of Prohibition. Catalyzed by the "Judgment of Paris" in 1976, in which Napa Valley labels defeated famous French wines, the region began to develop an international identity for its wines. Grape production expanded, as most agricultural lands were converted to vineyards. Agricultural water use began to increase, as technology became available to pump groundwater for spring frost protection and summer irrigation, leading to the construction of over 2,500 small reservoirs throughout the watershed.[38]

The rapid changes to the valley over the past two centuries have created environmental challenges, including the decline of local fisheries and the listing of the Napa River as impaired by sedimentation under the Clean Water Act. In response, there has been a growing interest in revitalizing the ecological systems that underlie the valley's health and productivity. A number of innovative efforts are currently under way, including the Napa River Flood Protection Project, which is guided by Living River principles;[39] Napa Green certification for sustainable winemaking; and the integrated river restoration by farmers in the Rutherford Reach Restoration Project. This Atlas was developed at the behest of a broad array of organizations involved in these and related projects—including Friends of the Napa River, Napa Valley Vintners, the Napa County Resource Conservation District, the Napa County Flood Control and Water Conservation District, the Napa Watershed Information Center and Conservancy, the Regional Water Quality Control Board, and the California State Coastal Conservancy—to support the ongoing effort to improve the ecological health of the valley.

To learn how the valley used to work and how it has changed, we turn to the documents produced along the way. At the same time that people were re-imagining and altering the landscape, they were also documenting it. Inadvertently, as part of their job, or as conscious chroniclers of their personal experience, these men and women measured, photographed, transcribed, painted, and sketched the characteristics of the valley. Taken together, the historical record is surprisingly detailed and voluminous. Historical ecology is made of these works: a story composed of hundreds of independent entries, linked by a shared landscape.

Timeline of data sources. (following pages) Useful documents include textual accounts, paintings and drawings, aerial and landscape photography, ecological collections, and technical reports. These materials were produced over many years by a wide range of people—a few of whom are highlighted here.

1800   1850

19th-century textual accounts

Land-grant materials

Landscape photographs

Paintings and drawings

Altimira's Diary

In the summer of 1823, Father José Altimira explored the river valleys draining to San Pablo Bay to select the site of the eventual Mission San Francisco Solano de Sonoma. While the Sonoma Valley was chosen, Altimira noted "large groves of oak trees" in the Napa Valley and foreshadowed the valley's agricultural future as "land proper for the cultivation of the vine."[33]

Bartlett's Travels

New York City bookstore owner John Russell Bartlett was appointed as U.S. commissioner for the survey of the new Mexican-American border. While his tenure was controversial, he took the opportunity to tour the West and is now recognized for his copious descriptions of landscape and culture—including two productive days spent in the Napa Valley during the spring of 1852.

Illustrations of Napa County

Clarence L. Smith and Wallace W. Elliott produced this commercial atlas in 1878, including over 50 lithograph views of local farms and homes and accompanying descriptions.

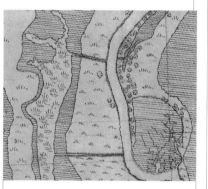

Mexican Land-Grant Sketches

As the mission system disintegrated, influential Mexican citizens submitted land-grant claims to the government, including a *diseño*—a sketch of the solicited property. While not as spatially precise as subsequent surveys, these drawings or watercolors are often highly descriptive, showing physical landmarks that served as boundaries, as well as notable natural resources such as creeks, springs, wetlands, and woodlands.

United States Coast Survey "T-sheets"

As part of the development of nautical charts for San Francisco Bay, 25-year-old David Kerr led a survey team through the tidal marshlands and surrounding valley as far north as the site of the present-day city of Napa during the summer of 1858.

General Land Office (GLO) Public Land Survey

A handful of local surveyors worked to establish the sectional grid of the federal Township and Range system, recording the size, location, and species of numerous "witness" trees as well as cultural and natural features crossed by the GLO transects.

Vischer Drawings

Artist Edward Vischer created a series of pencil sketches of Oak Knoll Farm during the 1860s, with a particular focus on valley oaks.

Land Case Testimony

Grantees and other early settlers later testified about the grant boundaries in American court proceedings, providing detailed discussion of the Napa Valley landscape at the time of the grants.

1900　　　　　　　　　　　　　　　　1950　　　　　　　　　　　　　　　　2000

Federal surveys

Aerial photography

Technical reports

Ecological records

USDA Soil Survey

Soil scientists E. J. Carpenter of the U.S. Department of Agriculture (USDA) and Stanley W. Cosby of the University of California described soil and drainage characteristics, water use, and agricultural patterns in a detailed map and accompanying report.

Early Aerial Photography

Over the course of several days during the summer of 1942, USDA pilots and photographers created the earliest comprehensive picture of the valley.

Contemporary Aerial Photography

Color digital ortho-photography made available by NAIP has a horizontal accuracy of 6 meters (20 feet).

Turrill and Miller Photographs

During 1906–1907, Charles Turrill of the photography firm Turrill and Miller conducted an extensive survey of the emerging wine industry in Napa and Sonoma valleys. He later explained, "I am fighting to preserve the history of California—for the accuracy of historic records."[34]

Willis Jepson

The great California botanist studied the Napa Valley flora throughout his lifetime, recording numerous observations in his field books. In 1934, at the age of 67, he wrote, "For pure delights and serene contentments, the floor of Napa Valley and its foothills impinging ruggedly on the plain, have for me great store of happy memories."[35]

Report on the 1955 Flood

After the devastating flood of 1955, engineer George Nolte proposed solutions for the Napa River in a technical report that also included a detailed map of the flood extent.

Modern Soil Survey

During 1965–73, soil scientists G. Lambert and J. Kashiwagi of the USDA created the modern soil survey of the valley.

A MAP OF THE NAPA VALLEY IN THE EARLY 1800s

These diverse sources can be integrated to show the character of the valley prior to significant Euro-American modification. Such a composite map is one of the most important products of a historical ecology study. This kind of ecological reconstruction shows us how the landscape supported native species in the recent past. When compared to contemporary maps, it reveals what has been lost, suggesting causes for species decline. An accurate historical ecology map can reveal how habitat patterns were controlled by physical factors such as soils, topography, and groundwater, guiding strategies for restoration. When compiled in a geographic information system (GIS), the historical map can be overlaid on contemporary imagery and other data sets to quantify local changes, discover habitat remnants, and identify restoration opportunities.

To make a map of the Napa Valley as it looked two centuries ago, we systematically compared and integrated several hundred independent documents. Different styles and eras of documents revealed different aspects of the landscape, filling gaps and enabling inter-calibration between documents to assess their accuracy and reliability.[40] By drawing upon these different vintages of information, the historical map becomes more than a snapshot view, but rather reflects a broader time frame. Using this array of sources, we assessed and mapped features that tend to be relatively persistent for decades, if not centuries (e.g., topographically controlled wetlands, mature oak stands, streams). The map is thus a composite picture representative of average conditions during the first decades of Euro-American settlement.[41]

During the period from which data are drawn, the landscape itself was of course being modified. However, the large-scale patterns of the valley landscape do not appear to have changed abruptly in the first decades

Certainty levels. To record the inevitable variations in source data and confidence, we used an attribution system to track sources and certainty levels for each mapped feature.[42]

Certainty Level	Interpretation	Size	Location
High/ "Definite"	Feature definitely present before Euro-American modification	Mapped feature expected to be 90%–110% of actual feature size	Expected maximum horizontal displacement less than 150 feet
Medium/ "Probable"	Feature probably present before Euro-American modification	Mapped feature expected to be 50%–200% of actual feature size	Expected maximum horizontal displacement less than 600 feet
Low/ "Possible"	Feature possibly present before Euro-American modification	Mapped feature expected to be 25%–400% of actual feature size	Expected maximum horizontal displacement less than 1,650 feet

of Euro-American management, but rather to have changed as land use shifted and intensified through the 19th and 20th centuries. Mission and rancho era sources (early textual accounts, *diseños*, and land-grant testimony) describe similar characteristics as the more detailed American sources produced in the following decades. The consistency of 19th-century historical records, as well as the lack of discussion of major changes in land-grant testimony or other early local accounts, suggests that most valley habitats were impacted more gradually. In fact, as is documented here, some characteristics persisted for surprisingly long— and are in some cases still evident today.

On a longer time frame, this 19th-century landscape was itself a temporary scene in a changing tableau. Landscapes are always evolving in response to climatic changes, as well as catastrophic events such as earthquakes and fires. The extent of different Napa Valley habitats shifted in response to wet years and droughts, as well as longer-term climatic patterns, including the little Ice Age (~A.D. 1450–1800) and the relatively stable climatic conditions of the past 100 to 200 years.[43] The map of early-19th-century Napa Valley is important not because it was a permanent or an ideal landscape, but because it represents the recent product of the physical, cultural, and ecological processes shaping the valley. As such, it is a tool to help us understand those processes and how they might work in the future.

Mapped habitat types. We found historical evidence for a variety of habitats in early Napa Valley. In addition to these features, we mapped streams as linear features.

Habitat Type	Description
Deep Bay/Channel	Tidal waters deeper than 18 feet below Mean Lower Low Water (MLLW)[44]
Shallow Bay/Channel	Tidal waters between MLLW and 18 feet below MLLW[44]
Tidal Flat	Intertidal areas of bare (less than 10% vascular vegetation cover, other than eelgrass) clay and silt, sand, and/or shell hash above MLLW[44]
Tidal Marsh	Intertidal areas that support at least 10% cover of vascular vegetation adapted to intertidal conditions[44]
Wet Meadow	Temporarily or seasonally flooded herbaceous communities characterized by poorly drained, clay-rich soils[45]
Alkali Meadow	Temporarily or seasonally flooded herbaceous communities characterized by poorly drained, clay-rich soils with a high residual salt content[46]
Vernal Pool Complex	Areas with seasonally flooded depressions associated with impervious subsoils and native vernal pool plant species[44]
Valley Freshwater Marsh	Freshwater, nontidal wetlands dominated by persistent emergent vegetation; temporarily to permanently flooded, permanently saturated[47]
Perennial Freshwater Pond	Permanently flooded, nonvegetated depressions[48]
Broad Riparian Forest	Wide (greater than 330 feet on a side) forested areas adjacent to the Napa River and/or its side channels[49]
Braided Channel	Broad stream channels (wider than 200 feet) characterized by largely unvegetated gravel/sand surfaces and limited woody vegetation[50]
Grassland/Wildflower Field	Low, herbaceous communities occupying well-drained soils and composed of native wildflowers and grasses[51]
Valley Oak Savanna	Valley oak–dominated communities with canopy cover less than 25% and a low, herbaceous understory[52]

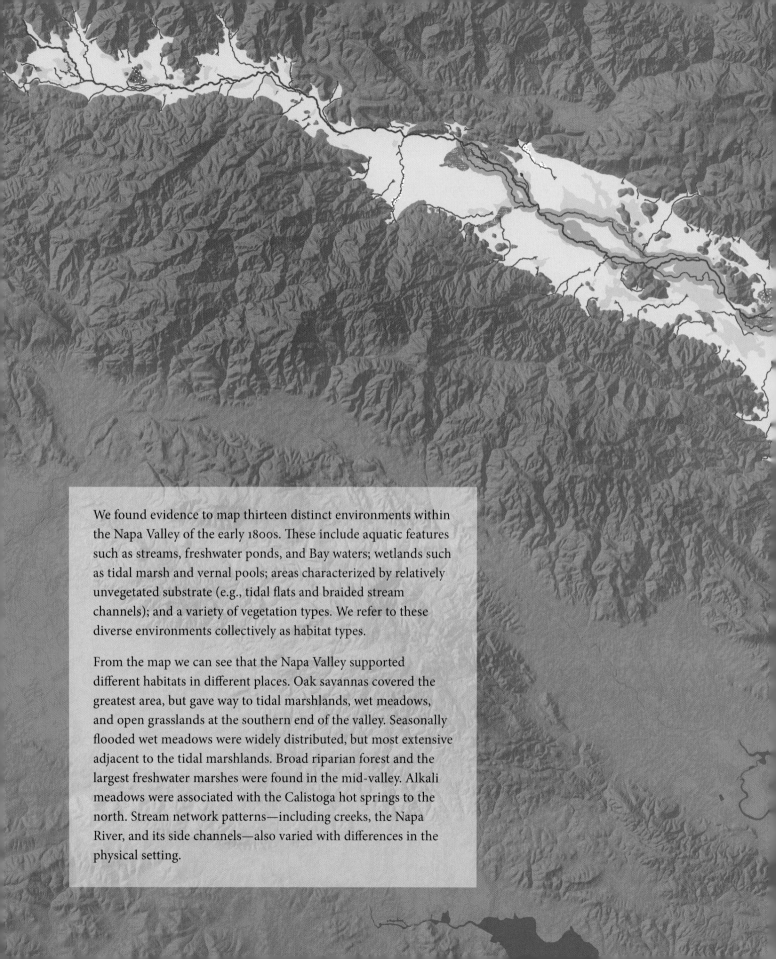

We found evidence to map thirteen distinct environments within the Napa Valley of the early 1800s. These include aquatic features such as streams, freshwater ponds, and Bay waters; wetlands such as tidal marsh and vernal pools; areas characterized by relatively unvegetated substrate (e.g., tidal flats and braided stream channels); and a variety of vegetation types. We refer to these diverse environments collectively as habitat types.

From the map we can see that the Napa Valley supported different habitats in different places. Oak savannas covered the greatest area, but gave way to tidal marshlands, wet meadows, and open grasslands at the southern end of the valley. Seasonally flooded wet meadows were widely distributed, but most extensive adjacent to the tidal marshlands. Broad riparian forest and the largest freshwater marshes were found in the mid-valley. Alkali meadows were associated with the Calistoga hot springs to the north. Stream network patterns—including creeks, the Napa River, and its side channels—also varied with differences in the physical setting.

The Napa Valley in the early 1800s. The map reconstruction integrates historical and contemporary information to illustrate landscape patterns prior to significant Euro-American modification.

Valley oaks at Calistoga. In this pair of stereographic images made around 1880, photographer Carleton Watkins captured the peculiar combination of large trees and open meadows that defined Napa Valley's oak savanna. Taken from slightly different positions, the images were designed for a special 3-D viewer. The foreground shows a portion of the Calistoga hot springs resort (misidentified as Sonoma County), including young ornamental trees and one of the resort bungalows (with horse and driver in front). *Watkins Stereo Collection no. 1792, courtesy of the California State Library.*

2 • OAK SAVANNAS AND WILDFLOWER FIELDS

View from the Calistoga Hotel, Sonoma Co., Cal. 1590.

Photographic Views of California, Oregon, and the Pacific Coast generally—embracing Yo Semite, Big Trees, Geysers, Mount Shasta, Mining, City, etc., etc. Views made to order in any part of the State or Coast.

The magnificent oaks are one great secret of Napa's beauty. Their rustling leaves and finely formed tops are the glory of the landscape scenery, and they everywhere, single and in groups, are scattered over the valleys.

—SMITH AND ELLIOTT'S 1878 *ILLUSTRATIONS OF NAPA COUNTY*

These splendid trees give the valley floors a distinction which is entirely lacking to them when the oaks are cut away to make room for orchards and gardens. It is however quite possible to offset largely the effects of this removal by using the Valley Oak on the main and cross roads as a roadside tree, particularly by planting at irregular intervals.

—BOTANIST WILLIS JEPSON IN 1909

For the many 19th-century travelers who visited Napa Valley, one element particularly captured the imagination: the great oaks. The trees inspired a substantial literature of superlative accounts in journals, books, and newspaper articles. Described by author Robert Louis Stevenson and botanist William Brewer, photographed by Carleton Watkins and Eadweard Muybridge, Napa Valley's giant oaks lent a sense of grandeur and beauty to the landscape that was widely appreciated, even as the trees declined.

The trees formed a complex savanna landscape of widely spaced individuals, occasional groves, and open meadows. The age of some of the massive trees (as much as half a millennium) suggested an ancient, carefully designed landscape that many likened to an English park. John Russell Bartlett, perhaps the best of the oak chroniclers, described how "this romantic valley…answers to the idea one has of the old and highly cultivated parks of England, where taste and money have been lavished with an unsparing hand, through many generations."[1]

Occupying the fertile loam soils of the alluvial fans and floodplains, the oak lands soon became the famous agricultural lands of Napa Valley. Yet they had been working landscapes for centuries, managed by native tribes through prescribed burns to generate prolific acorns, native grains, wildflower seeds, and edible bulbs. The oaks played a central role in the ecology of the valley, supporting dozens of native wildlife species, from the acorn woodpecker and white-breasted nuthatch to the wood duck, Pacific pallid bat, and grizzly bear. Between the trees, herds of pronghorn and tule elk traversed the open meadows, foraging on native grasses and wildflowers.

Napa is shaded by an oak grove not yet demolished. The fine old trees stand quite thick on the south side of the town, and, as in most parts of the valley, have been preserved with remarkable care and good taste.

—SACRAMENTO DAILY UNION, 1860

The first decades of Spanish and Mexican land use did not immediately threaten the valley's oaks. Ranchers valued the trees for the shade they provided during the long hot summer. Underneath the oak canopies, cows found green grass when pastures had gone dry.

Early American activities also did not necessarily conflict with the oaks. Settlements took advantage of the oak groves, for both their aesthetic value and the practical benefit of shade during the hot summer. Ranching and grain farming—the predominant activities through the 1860s—often coexisted with the dispersed oaks and in fact benefited from them, as the *Daily Alta California* described in 1860: "The shade…protects the land against the burning sun, and preserves its moisture. The trees also keep off the frost from the grain and soil. Fruit, as well as grain, grows better under large trees than when unprotected."[2] Providing only poor-quality timber, valley oaks were often more valuable standing. While clearing and firewood cutting caused local impacts, and ranching and grain farming likely reduced the survival of new seedlings, the oak lands appear to have persisted largely intact until the 1880s.[3]

Over the course of just a few decades, however, the rapid spread of vineyards and orchards altered the landscape. These intensive, higher-value plantings required the sunlight intercepted by broad oak canopies; over 100 trees per acre left little room for anything else.[4] By 1910, farmers had clear-cut most of the oak savannas.[5]

This rapid transformation took place prior to any living memory: the 19th-century travelogues recorded a landscape no one has seen for over a century. Yet despite this cultural and ecological discontinuity, significant elements of the oak lands could still be recovered today. In fact, the natural savanna pattern lends itself to coexistence with developed landscapes. The understory grasses and wildflowers, while now mostly excluded from the valley, could also return in places if desired.

This chapter establishes the characteristics of the lost and forgotten savanna landscape. We start by investigating the spatial patterns of the savannas and decoding the evidence for their composition. We then explore the herbaceous understory: the open grasslands and wildflower fields. The final spreads consider the cultural significance of the dominant savanna tree—the valley oak—and its potential role in the future valley.

Valley oaks within the modern landscape. A mature tree at Trefethen Family Vineyards (top) and a mixed-age stand along a farm road (bottom). *Photographs by Susan Schwartzenberg, April 18, 2009 (top) and November 23, 2008 (bottom).*

Oak canopy. The massive valley oak structure (far left) creates an arboreal world suspended several stories above the realm of human activity. Oak canopies sing and flutter with activity, particularly when part of a larger savanna network. The acorn woodpecker (left) lives in complex family groups in oak canopies; these birds store acorns for the winter in granary trees. *Far left: photograph by Susan Schwartzenberg, April 18, 2009. Left: photograph by David Hoffman, courtesy of CalPhoto.*

"The Oaks." (above) Some of Napa Valley's oak groves were particularly well documented, such as the group shown in this postcard. The original Turrill and Miller photograph, taken in April 1907, was hand-tinted to create the color postcard image. A few of these trees still remain in a mobile home park on the west side of Calistoga. *Courtesy of The Bancroft Library, UC Berkeley.*

Wheat harvested underneath valley oaks, 1907. (left) The view is from Berryessa Valley, the valley to the east of the Napa Valley, where grain farming continued longer. The coexistence of valley oaks and grain was described in Napa Valley a half century earlier: "The roots of these trees strike into the ground so deep that the plow never touches them, though running as close as a team can go to the trunk, and the grain stands as thick, and bears as heavy a crop within a foot of the trees, as on any portion of the ground."[6] *Photograph by Turrill and Miller, courtesy of the Society of California Pioneers, catalog no. C028036.*

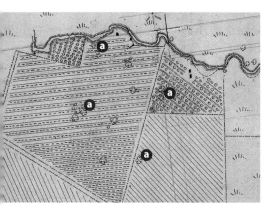

Oaks in maps. A *diseño* for the Carne Humana land grant, circa 1840 (top), shows a *roblar* (oak grove), probably at St. Helena on the Sulphur Creek fan. Oaks ⓐ were often preserved in early agricultural fields, as shown in the 1858 U.S. Coast Survey (bottom). *Top: courtesy of The Bancroft Library, UC Berkeley. Bottom: courtesy of NOAA.*

MAPPING THE OAK LANDS

Despite the abundance of inspired textual description, no one mapped the overall distribution of oaks in the valley. Some visitors emphasized their ubiquity, reporting that "all the lower valley lands are dotted with huge oaks"[7] or describing how "oaks…dot the lawns and fields on every side and shade the road."[8] Dramatic descriptions can be misleading, however. In the similarly described Santa Clara Valley, actual oak lands covered about 60% of the alluvial valley (not including the tidal lands).[9] As we consider their past and future place in the Napa Valley, we would like to know exactly where they were found.

To document historical distribution, we compiled a wide range of evidence, ranging from living "heritage" oaks and remnant trees visible in 1942 aerial photography to 19th-century maps, texts, lithographs, and landscape photographs. The resulting data set comprises over 3,000 points describing the historical extent of the oak lands.

Together, these data suggest that oaks were, in fact, quite ubiquitous. Prominent stands were documented at the northern end of the valley in Calistoga, as far south as Suscol Creek (several miles south of Napa), and at numerous sites in between. Evidence for individual trees occurs widely through the valley.

Integrating these data, we estimated 36,000 acres of oak lands, covering 60% of the valley above tidal extent (a similar proportion to the Santa Clara Valley). We distinguished areas of greater and lesser certainty. The areas without direct evidence for oaks were typically covered by intensive agriculture at the time of early aerial photography. Where 19th-century materials likely to represent oaks (e.g., GLO surveys, the 1858 T-sheet, landscape photography) are available, they almost invariably show oaks. Accordingly, in addition to the high-certainty oak lands, we mapped adjacent areas that were intensively cultivated by the 1940s and had suitable soils, identifying them as probable oak lands.[10]

Not all parts of the valley supported the trees, however. Oaks were consistently excluded from the heavier, more compact soils of tidal marshland, freshwater wetlands, and wet meadows. Heavy clay subsoil similarly kept trees out of the broad Carneros plains, which instead supported open meadows.[11] Oaks grew in the loam soils of Napa Valley's alluvial fans and floodplains, benefiting from the same good drainage and access to groundwater that makes these areas prime agricultural lands.

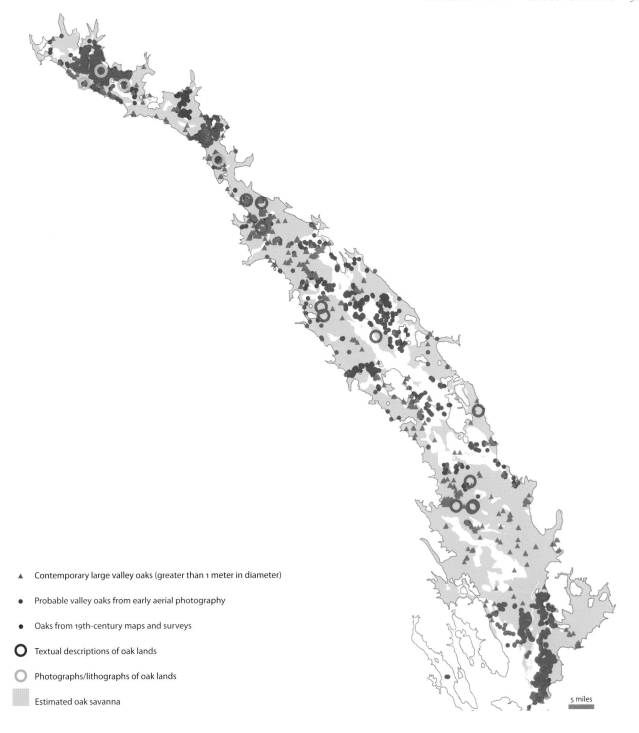

- ▲ Contemporary large valley oaks (greater than 1 meter in diameter)
- • Probable valley oaks from early aerial photography
- • Oaks from 19th-century maps and surveys
- ⊙ Textual descriptions of oak lands
- ⊙ Photographs/lithographs of oak lands
- ▨ Estimated oak savanna

5 miles

Evidence for the historical distribution of oak savanna. Data describing individual trees include large contemporary valley oaks,[12] probable valley oaks mapped from 1942 aerial photography,[13] probable oak savanna trees from detailed local maps produced between 1858 and 1873,[14] and witness trees identified as oaks by GLO and county surveys. Evidence for groups of trees includes the section descriptions of the GLO surveys (e.g., "scattered trees"), lithographic views of individual farms and properties, landscape photographs, and textual accounts.[15] These data identify areas of well-documented oak presence and provide a basis for estimating the broader probable extent.

SCATTERING OAKS

The trees do not cover the ground thickly, but are scattered here and there...just far enough apart to make them imposing.

—MARY CONE, 1876

Based on the prominence of oaks in historical accounts, some writers concluded that the oaks formed a dense forest canopy. In 1928, author W. P. Bartlett presumed that, before clearing, "Napa Valley was an oak forest, under which one could walk in dense shade." But while oaks were the dominant visual element in most parts of the valley, earlier accounts reveal a relatively open pattern of light and shade: "scattered" or "scattering" oaks, a valley "dotted" with trees.[16] These phrases, as well as early maps and photographs, describe an incomplete canopy cover corresponding to what we would today call a savanna.[17]

Seventy-four years earlier, a different Bartlett, John, recognized the almost paradoxical pattern of giant trees that dominated the landscape, yet took up relatively little space. The valley was "studded with gigantic oaks, some of them evergreen, though not so close together as to render it necessary to cut away to prepare the land for cultivation." Within the savannas, trees occurred in repeating, finer-scale groupings, leaving large open areas: "These magnificent oaks are found sometimes in long lines, and again in clusters of twenty or thirty, forming beautiful groves; then again a space of ten or twenty acres will occur without a single tree."[18]

The pattern of groves, glades, and individual trees was described similarly by independent observers. Smith and Elliott reported, "they are disposed about the plain in most lovely groups, masses or single ones"; Stevenson described how "a great variety of oak stood, now severally, now in a becoming grove, among the fields and vineyards."[19] As much as the trees themselves, it was this intricate, landscape-level spatial pattern that fascinated observers, who remarked on a scene "as perfect as if the work of art" or, alternatively, "arranged by Nature with such exquisite symmetry as art could never accomplish."[20]

Groves and glades. This view of Knights Valley (located north of the Napa Valley, on the opposite side of Mt. St. Helena) by Virgil Williams could also illustrate conditions in the nearby Napa Valley, where he spent much time. Clumps of oaks are interspersed with wildflower fields. Yellow patches in the oak openings may be goldfields or tidy-tips. *Courtesy of the Oakland Museum of California.*

OAK SAVANNAS AND WILDFLOWER FIELDS • 33

"Group of Oaks (White Oak) in Napa Valley." (above) Cows shelter under scattered oaks in a drawing made by Edward Vischer in the 1860s. *Courtesy of the Edward Vischer Collection, Special Collections, Honnold/Mudd Library, Claremont University Consortium.*

Oak patterns on the Mill Creek fan. (left) A 1942 aerial image shows varied patterns of oak spacing before clearing, just north of St. Helena. Red box on map below shows location.

500 feet

BURNING THE VALLEY

Except in the mountains there is no heavy timbered land in the county. The land when fenced is all ready for the plow.
—C. A. MENEFEE, 1873

Through parks of evergreen oaks that looked as perfect as if the work of art…
—SAMUEL BOWLES, 1865

The savanna character of the valley was due not just to the grand trees but also to the low vegetation between and beneath them. Individual trees stood out on open plains easily viewed and traversed by people. Many visitors described the tidy, well-maintained valley as park-like, while others thought it appeared "as if arranged by the most skillful landscape gardener."[21]

The "skillful landscape gardener" invoked by visitors was in fact accurate: the savannas were shaped by human hands. Bartlett speculated that the lack of undergrowth might have been due to "fire since the occupation of the country by the Spaniards,"[22] but in fact the timing was the reverse. Long-term indigenous burning of the landscape was likely most responsible for its open character.

Ethnographic and ecological research throughout California has documented the intentional use of repeated, low-intensity fires by native peoples to maintain open meadows and savannas.[23] Napa Valley tribes used these same practices. In fact, the first detailed description of the region includes an account of indigenous pyro-management, as Altimira encountered a recently burned meadow while approaching the valley from Sonoma in 1823.[24]

Napa Valley tribes used fire to promote edible native grasses and wildflowers, improve soil fertility, catch insects and small game, keep the valley open for hunting and movement, and reduce the risk of catastrophic fire.[25] They continued these practices even after the establishment of the Sonoma Mission, as indicated by the 1836 treaty signed between Lieutenant Vallejo and neighboring tribes to control their burning.[26] Land management with fire was an essential part of designing and maintaining the valley over time, such that the term for the month of June in the Wappo language (as documented in the Alexander Valley, about 10 miles to the northwest) translated as "burn-the-valley moon."[27]

Indigenous land managers shaped the Napa Valley's characteristic savannas and meadows, using controlled burns to keep shrubs and trees from gradually invading the valley.[28] They created habitat for the abundant elk, pronghorn antelope, and deer reported by visitors.[29] These practices also had a consequential and unpredictable historical implication. They created the fertile, open landscape that was so easily converted to Spanish ranches and American fields.

A Wappo elder. The Caymus, Canijolmano, and Mayacma bands were part of the larger Wappo-speaking people. Edward S. Curtis took this photograph in 1924 as part of his series on the North American Indian. Though criticized for photographically presenting Indians in staged scenes to match the stereotypes of the time, he took more than 40,000 photographic images of 80 tribes, in addition to recording extensive information about native culture. While Wappos were nearly wiped out in the early 1800s, many still live in the region today, particularly in the Alexander Valley. *Courtesy of the Northwestern University Library.*

"Skeleton of a White Oak and Oak Grove Vistas in Napa Valley." Ranching and farming occupied a landscape shaped by centuries of indigenous management. Ancient trees, such as the "skeleton" in this 1864 drawing by Edward Vischer, could be more than 500 years old, having germinated in the Middle Ages. *Courtesy of the Edward Vischer Collection, Special Collections, Honnold/Mudd Library, Claremont University Consortium.*

A live oak persists in a vineyard south of Calistoga, April 10, 1907. This photograph was taken 30 years after journalist Benjamin Parke Avery eloquently described the dark green, densely leafed live oaks: "evergreen oaks, with their short trunks, cauliflower-shaped masses of intensely dark green foliage, and sharp shadows, then seem like oases in the hot expanse—grateful islets of verdure in a sea of shimmering yellow light."[38] *Photograph by Turrill and Miller, courtesy of the Society for California Pioneers, catalog no. C013325.*

SAVANNA COMPOSITION

Several lines of evidence indicate that valley oaks were the predominant component of Napa Valley's oak savannas. Of the 48 witness trees identified as oaks in land surveys, 31 (65%) were white oaks, as valley oak was often called.[30] Most trees visible in landscape photographs of oak groves are valley oaks. Valley oaks were also described most frequently in textual accounts, referred to alternately as white oak, burr oak (a similar-looking white oak of the eastern and midwestern U.S.), willow-oak, or valley oak.[31]

Other trees were important components of the savanna. Witness tree records suggest that the deciduous California black oak was the second most common oak in the valley (27%); several textual accounts affirm the species' presence.[32] Coast live oaks (named for their year-round foliage) were also identified as a significant component of the savanna, although there is some inconsistency in the descriptions, perhaps suggesting regional variation in savanna composition. Bartlett described the valley as "studded with gigantic oaks, some of them evergreen."[33] Several observers seemed to suggest greater prevalence,[34] but live oaks accounted for only four of the 48 witness tree oaks. In the most famous savannas, such as the Oak Knoll area and Calistoga, live oaks were a minor component, as traveler Mary Cone noted: "Occasionally a live-oak is seen among them, which, being much less grand and beautiful, looks as though it might be glad to dwell in such grand company."[35] In addition to oaks, there is limited evidence for other tree species, particularly in the upper valley, where Bartlett stated that "the valley presented a greater variety of trees" and the *Daily Alta California* emphasized the increased abundance of firs.[36]

Jepson reported remnants of ponderosa pine and gray pine, while other sources suggest the presence of a variety of conifers, particularly at the valley margins.[37] These trees were more common in the adjacent hills and canyons, but they likely extended onto the valley floor in places. Providing much more valuable timber than oaks, any conifers present on the valley would likely have been subject to rapid cutting, reducing their presence relative to the oaks.

A white oak on the margin of a prune orchard, October 11, 1906. Another Turrill and Miller photograph illustrates Avery's description of the contrasting shape and texture of valley oaks: "the massive trunks, tall forms, and expansive foliage of the deciduous oaks, present a striking contrast to these hardy dwarfs who have to struggle for life." He also noted the hanging tendrils characteristic of a mature tree: "The willow-oak, remarkable for the pendant strips of leafage nearly touching the ground, from which it derives its name, is particularly conspicuous."[39] *Photograph courtesy of the Society for California Pioneers, catalog no. C014054.*

A WORLD OF WILDFLOWERS

The low understory vegetation of oak savannas was so rapidly invaded by European plants that its pre-contact composition is not obvious. Deciphering the history of these herbaceous communities has challenged California researchers for decades. Recent studies have emphasized that wildflowers, not just grasses, were a major component.[40]

In the Napa Valley, historical accounts describe extensive fields of wildflowers, as well as native grasses. While we do not know the exact composition of the herbaceous cover beneath and between the scattered trees, wildflower fields clearly represented one of the prominent features of the valley.

The first waves of bloom typically appeared in March, following the winter rains, as witnessed by Bartlett in the spring of 1852 near American Canyon: "Wild flowers of varied hues were thickly scattered around, and everything showed that the heavy and continued rains had given new life to vegetation." As he headed north from Oak Knoll the next day, he noted the intermixing with grasses: "A luxuriant growth of grass, studded with brilliant wild flowers, lined our path."[41] Carpets of flowers imbued the landscape with not just color, but also fragrance. A train ride passed through "the beautiful Napa Valley, surrounded by a world of wild flowers of most gorgeous hues, which covered the plains on either side, and loaded the air with a rich perfume."[42]

One of the striking wildflowers was the famous California poppy, which was said to have attracted pioneer George Yount to the valley, and whose "deep orange tint" was conspicuous "in the oak-openings."[43] Jepson immersed himself in the complexity of the colorful patchwork, recollecting "a radiant springtime when for four glorious weeks I botanized the whole length of the valley….Cream cups, poppies, clovers, collinsias, and owls clover filled the meadows and fields."[44]

The loss of this remarkable expression of local climate and ecology—equal in seasonal drama to the fall colors of the East Coast—is a largely overlooked transformation of the landscape.[45] Despite the replacement of native wildflowers on the valley floor by cultivated areas and invasive plants, the native seeds can remain dormant in the soil for decades, suddenly reappearing in profusion after years of absence when conditions allow them to germinate.[46] Native wildflower seed banks lie intact in places, waiting for the right conditions to return.

Local color. Napa Valley's native wildflower fields produced a seasonal, multicolored display, changing by the week as different species bloomed and faded. Some patches persisted long enough to still be remembered today, such as Inez Pometta's "treasured fields" of lupine near Oakville Grade and Highway 29.[47] Occasionally, blooms return in vineyards and fields. *Photographs of tidy-tips (top) and Parry's tarplant (bottom) by Jake Ruygt, April and July 2008.*

Napa wildflowers in postcards. California poppies, purple owl's clover, and lupine in a circa 1950 postcard (opposite page, top). Lupines in a circa 1900 scene titled "The orchards at Calistoga are veritable flower gardens in April" (opposite page, bottom). *Top: courtesy of Todd Schulman. Bottom: courtesy of Sharon Cisco Graham.*

VALLEY OAKS AND PEOPLE

The particular charisma of the oak lands derived largely from a single species of tree: the stately valley oak. Native only to California, the species is the largest North American oak and one of the longest lived, commonly spanning three or four centuries. Valley oaks dominated many of California's fertile valleys at the time of European contact, so much so that historian John S. Hittell identified them as "one of the most important and characteristic features of the California landscape."[48]

This uniquely Californian landscape ironically provided a compelling link to the familiar home landscapes of newcomers. Some compared the savanna with elegant English parks;[49] others were reminded of the South, as in an 1873 description: "huge oaks, with pensile limbs like trailing grapevines, which fairly sweep the ground, and often loaded with greenish-gray moss, which gives the landscape such an aspect as that of the lowland Country of Texas and Louisiana, where the Creole-moss abounds."[50] Valley oaks were also commonly identified with elms of the Midwest and East Coast, having "the pendant grace seen in the Eastern states only in the elm."[51]

The presence of valley oaks made Napa Valley appealing and habitable to Euro-American settlers. Distinctly Californian, the trees nevertheless provided the classic savanna setting valued by people around the world.[52] Not coincidentally, early accounts described how settlements were designed within the savannas, to take advantage of the trees. Thompson stated in 1857 that "the scattered oak trees and groves afford tasteful sites for residences," while the *Sacramento Daily Union* reported three years later that "Napa is shaded by an oak grove not yet demolished."[53]

In 1878, Smith and Elliott described how "the pretty little village of St. Helena, with its 50 or more houses, many of them neat and white, nestled among grand old oaks, was very picturesque"—a description echoed by St. Helena resident Babe Learned, recollecting her childhood four decades later: "We could walk around town under shade…everyone had an oak in their yard."[54] Commercial complexes utilized the oaks as well: at Charles Krug's winery, built within a grove, the trees were "most useful in the protection they afford from the summer heat."[55] The inherited architecture of valley oaks sheltered growing Napa Valley towns.

Under the oak canopy. Roadside oaks provide shade on a slow, hot ride down the valley from St. Helena to Napa in this image from the fall of 1906. The horse team huddles under the canopy while the driver reclines against the massive trunk. *Photograph by Turrill and Miller, courtesy of the Society of California Pioneers, catalog no. C014054.*

OAK TRAJECTORIES

Despite the early integration with towns and farms, the overall oak population within the valley has declined precipitously. Most of the loss took place a century ago with the development of orchards and vineyards, but the number of mature valley oaks has continued to fall during the past 70 years, raising questions about the long-term fate of this significant cultural and ecological component of the valley.

At the same time, acorns continue to germinate in yards and in untended areas along fences and roads; young valley oaks are not uncommon. Occasional large trees can still be found shading houses, business courtyards, parks, and roadways. The following three pages look at different oak trajectories as expressed at three sites, each located within a mile of Oak Knoll: the preservation of a mature oak grove, the persistence of roadside savanna trees, and the expansion of a new, modern oak grove.

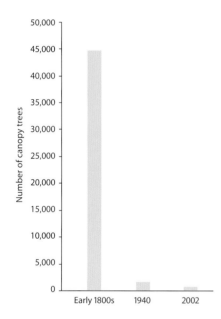

Valley oak decline. (left) This graph compares estimates of the total number of mature canopy trees in the valley at three points in time. The majority of loss occurred by 1942, but decline has continued. The recent total is based on a census of large trees conducted in 2002. The 1942 total includes the contemporary trees and additional similarly sized trees, mapped from early aerial photography. For the early-1800s estimate, we multiplied the total mapped area of oak savanna by density estimates from aerial photography and early maps.[56]

Twentieth-century oak removal. (below) While pockets of oak savanna survived into the 1940s, many have since been removed.

October 1906

October 2001

Grove of white oaks. (top, bottom right) The prominent oaks on the Dry Creek fan at Oak Knoll were first noted by Bartlett in 1852, when he described encountering "a new aspect" of the valley at Joseph Osborne's house. They were Napa Valley's most famous oaks, appearing frequently in photographs and accounts—especially the grove at Osborne's Oak Knoll Farm. In 1858, a California State Senate report commented on "the preservation of the magnificent native oaks" at Oak Knoll, emphasizing that "these trees, once cut down, cannot be replaced in many generations, while, if saved, they are at once, and perpetually, an ornament and a blessing." In fact, the stand at Osborne's Farm, perhaps the valley's emblematic oak grove, has been largely maintained to date, while every other significant stand in the valley has been removed. The grove does show signs of decline: there are less than half as many trunks in the contemporary image, and the trees have sparser canopies. But it remains the best living example of the valley's historical oak groves. The large, round-canopied tree closest to the camera appears to no longer be present, but the distinctive leaning trunk in the center maintains the same angle, with a century more girth. *Top: photograph by Turrill and Miller, 1906, courtesy of the Society of California Pioneers, catalog no. C027512.*

"View on Oak Knoll Farm" circa 1866. (bottom left) While valley oak does not produce commercial lumber, branches at the famous grove were used to construct ornate oak furniture beneath the oak overstory. *Photograph by Lawrence and Houseworth, courtesy of The Bancroft Library, UC Berkeley.*

The mystery of seven big trees. Driving along Highway 29, one passes roadside plantings of oleander, bottlebrush, eucalyptus, and other introduced species, along with occasional groups of native oaks. One of the most striking is a series of giant valley oaks just north of Oak Knoll Avenue. The trees are so well aligned to the road, with similar size and spacing, that at first glance they seem like the other plantings. But the size of the trees—trunks 41 to 49 inches in diameter—suggests that they must be at least 150 to 250 years old. The trees sprouted from acorns before significant clearing, when there were thousands of trees in the valley. It is likely that this linear grove is actually a slice of the original savanna, the fortuitous few that happened to find themselves between road and field as their neighbors were cut down. Scattered through the valley, these just-missed trees infuse the present-day Napa Valley landscape with the native landscape of centuries past. Beyond their historical significance, these and other surviving trees are living links to the past landscape, contributing viable genetics and ecosystem functions, that could serve as nodes for future restoration of valley oak populations. *Photographs (above and bottom left) by Susan Schwartzenberg, November 23, 2008. Photograph (bottom right) June 2003.*

Linear groves. In a few places, Napa Valley roadsides also demonstrate impressive regeneration of young trees descended from the former savanna. At the most well-developed sites, mixed-age oak communities can be found in thin lines along highway right-of-ways or farm roads. At these sites, local farmers have preserved mature trees and allowed their acorns, likely planted by scrub jays and squirrels, to develop in these narrow but productive corridors of ecological opportunity. As Jepson noted a century ago, these settings represent viable communities within the rectilinear framework of highly managed landscapes. *Photograph (above) by Ruth Askevold, November 23, 2008. Photograph (below) by Susan Schwartzenberg, November 23, 2008.*

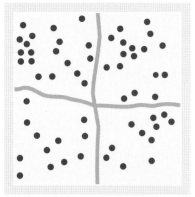

early 1800s

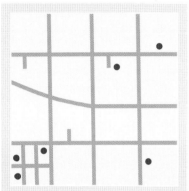

circa 1950

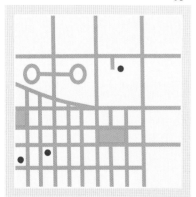

circa 2000

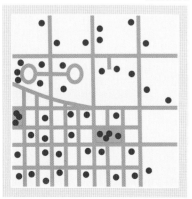

2050?

RE-OAKING THE VALLEY

These examples of persistence and recovery within the intensively developed landscape suggest the potential for reintegrating oaks more broadly within the valley. With their small footprints but large ecological effects, oaks and other native trees could be strategically incorporated in many places—in yards, parks, and parking lots; along roads, railroads, and fence lines. Such a strategy was advocated by Jepson, who, a full century ago, recognized this potential:

> The beauty of the interior and some coast valleys is very largely due to the presence of the scattered groves of the Valley or Weeping Oak. These splendid trees give the valley floors a distinction which is entirely lacking to them when the oaks are cut away to make room for orchards and gardens. It is however quite possible to offset largely the effects of this removal by using the Valley Oak on the main and cross roads as a roadside tree, particularly by planting at irregular intervals. A few owners of roadside frontage have made such plantings and there must be numerous others, permanent land owners, who at nominal expense would be glad to help restore, where destroyed, the park like beauty of the Californian valleys.[57]

As we anticipate changing climatic conditions in the future, "re-oaking" the valley would provide a range of valuable ecological services, such as shading and temperature modulation, carbon storage, and improved air quality. With the incorporation of evergreen live oaks, which capture and slowly release rainfall from their canopies, the new savannas can reduce the severity of runoff pulses.[58] Valley oaks occupy a broad climatic gradient and, once established, are relatively tolerant of arid conditions—particularly if groundwater levels are maintained within acceptable levels.

Napa Valley's oak savannas displayed a complex spatial pattern of widely spaced trees and more dense groves that facilitated ecological linkages. These landscape-level patterns may be adapted, as Jepson described, to the more rectilinear structure of the contemporary landscape to create ecological connectivity and functional populations of oak-associated birds, bats, and other species.[59] Establishing viable densities of valley oaks—whose distance of genetic exchange appears to be relatively small[60]—may help the tree persist in the face of climate change. Having declined for a century and a half, this icon of the valley may have a larger role to play in coming decades.

Conceptual model of past and future oak savanna trajectory. For most of the 19th century, oak savannas occurred broadly throughout the valley. Dirt roads went around trees and, for the most part, so did ranching and early agriculture. By the 1940s, most of the valley had been cleared for orchards, vineyards, and cities, but a few trees remained in pasture lands, along roads, and as shade trees in towns and on farms. Despite some preservation of existing trees, oak decline has continued during the second half of the 20th century. In the future, densities and patterns similar to historical conditions could be achieved through strategic planting and stewardship along roads and fence lines, in parks and yards, and as part of commercial landscaping.

Native landscaping. (left) Gardeners and landscapers have begun to draw upon the repertoire of native grasses, wildflowers, and trees for drought-tolerant landscaping. Even in relatively small places, this indigenous palette gives a glimpse of the color and texture of the native Napa Valley landscape, while providing habitat for bees, bats, and birds. This young valley oak with understory wildflowers is part of the parking lot landscaping at the DoubleTree Napa Valley hotel. *Photograph April 18, 2009.*

Oak orchard. (below) These valley oaks established themselves in an untended Calistoga prune orchard, germinating in the uncultivated area between tree rows. The orchard was cleared and the oaks were left. The flourishing young grove still maintains an unusually orderly pattern—the negative image of the vanished orchard. These accidental restorations show the resilience of Napa Valley's oaks and the continuing suitability of conditions. *Photograph November 6, 2005.*

Garnet Creek, near Calistoga. Some Napa Valley creeks were lush and densely wooded, while others were washy and open, like this summer-dry segment of Garnet Creek, shown in a historical postcard. The unstable nature of this reach necessitated the addition of a third arch to span the channel. Such streams were still appreciated: the postcard (mailed by a local resident in 1911) begins, "Dear Jack—I wonder if you remember this bridge where you used to go sometimes and throw stones in water…. Your Loving Aunt Mabel." *Courtesy of Sharon Cisco Graham.*

3 · CREEKS

...through which bubbles a clear and limpid trout stream, whose silvery sheen can at times be caught sight of from the county road.

—LOCAL HISTORIAN CAMPBELL AUGUSTUS MENEFEE
DESCRIBING W. C. WATSON'S VINEYARD NEAR RUTHERFORD IN 1873[1]

Cutting through the oak savannas were sinuous corridors of aquatic habitat: the valley's narrow but consequential creeks. Occupying a fraction of the area of the oak lands, creeks nevertheless played a critical role in the valley landscape, as they still do today. The Napa Valley's creeks connect the hills to the valley, conveying the water and sediment needed to maintain alluvial soils, groundwater aquifers, valley wetlands, and riverine habitats.

Creeks support a high density and diversity of wildlife, providing an array of environments distinct from the surrounding land. Songbirds, newts and tree frogs, striped skunks and black-tailed deer all use creek corridors for various lifestages. Steelhead ascend the creeks each winter to their spawning grounds in the upper watershed; the Napa Valley creeks provide some of the best unblocked steelhead migration pathways in the Bay Area.[2] Often conspicuous as ribbons of dense forest within the open savanna landscape, the creeks serve as landmarks for property boundaries and land grants. People have valued the creeks as trout streams and sources of "everlasting" water.[3]

Many generations of indigenous, Spanish, Mexican, and American residents did not draw heavily on creek flows for water use. During the 19th century and much of the 20th century, orchards and vineyards were dry-farmed, so people diverted relatively small amounts of water for household use or small gardens. During this time, however, other impacts took place. Cattle grazed on riparian vegetation. In town and on farms, creeks began to be redesigned to improve the drainage characteristics of the valley. Many creek segments were completely filled in, while others were artificially extended or channelized to drain the lowlands.

Views of contemporary Napa Valley creeks. *Top: Dry Creek, August 2, 2010, by Jonathan Koehler. Bottom: a small creek in the Carneros district, June 20, 2010. Opposite page, top: Rector Creek, December 11, 2006. Middle: Kimball Creek, May 8, 2003, by Jonathan Koehler. Bottom: Tulucay Creek, April 9, 2007.*

Between 1946 and 1959, reservoirs were constructed on upper Conn, Rector, and Bell creeks to store water for local municipalities, removing access to extensive historical steelhead habitat and reducing stream flow to the valley. Numerous smaller reservoirs have been constructed on the valley floor to store water for frost protection and irrigation. There is growing concern that withdrawals from creeks and groundwater to fill storage ponds may reduce spring and summer flow, to the detriment of native fish populations.

Despite these impacts, Napa Valley's creeks maintain a prominent presence in the contemporary landscape. In fact, there are actually more miles of creek today than there were two centuries ago—although much of the network provides limited function. Diverse native fish assemblages still persist in many creeks (see pages 114–115). Riparian tree corridors have been constrained by adjacent land use, but have generally not been completely cleared. There is even evidence that riparian forest has returned in places, quickly responding when people have made room. One significant dam, on York Creek in St. Helena, has filled in with sediment and is being considered for removal to increase steelhead access to the upper watershed; incipient restoration efforts are underway on other tributaries. If provided sufficient space, Napa Valley creeks can still play their role as all-important corridors of energy, water, sediment, nutrients, and wildlife.

To explore some of the significant changes to the creeks, we first examine the discontinuous channel patterns and drainage challenges that led American settlers to redesign how water moves through the valley. In the second half of the chapter, we look at other characteristics of the creeks: riparian forest, channel incision, braided channels, and summer water.

A creek corridor through agricultural lands, 1858. (left) The cumulative length of Napa Valley's historical creeks exceeded the length of the Napa River by a factor of seven. Today, more than thirty named creeks flow through the valley and into the Napa River, as well as dozens of less prominent natural and artificial tributary channels. *Kerr 1858, courtesy of NOAA.*

I know of no stream between the 'ojo de agua' [spring] and Oakville that has a channel through to Napa River.

—NATHAN COOMBS, 1861

SPREADING STREAMS

We typically picture creeks as part of a continuous network, with channels emerging from canyons onto the valley, merging with adjacent tributaries, and becoming progressively larger until they reach the Bay or ocean. In the Napa Valley, as in much of the Bay Area, creek networks often did not follow this model.[4] Historical evidence indicates that while some larger creeks maintained channels across the valley and into the Napa River, many instead dissipated, spreading their flow broadly across the lowlands. Rather than being concentrated, stream flows were distributed and attenuated, reducing flood peaks downstream, recharging groundwater and wetlands, and depositing sediment to build and maintain the fertile valley.

Part of the reason the Napa Valley's creeks behaved differently than might be expected lies in the geologic foundation of the valley, which is fault-lined and uplifting, creating a sharp break in gradient as creeks exit the steep canyons onto their gently sloping alluvial fans. The change in slope and unconfined setting at the apex of the fans cause streams to slow down, spread out, and deposit coarse sediments. Water sinks quickly into the porous, gravelly fan soils. Creek channels jump their banks and change course relatively frequently, often resulting in a less well-defined channel.[5]

The Napa River itself also played a role in the discontinuous channel network. Alongside its channel, the river maintained a broad natural levee of higher ground that could be penetrated only by larger tributary creeks. Many of today's tributaries did not have sufficient power to cut channels across the lowlands and break through this ridge of sediment deposited by river floodwaters.[6]

Mid-19th-century maps illustrate these patterns, often showing creeks ending abruptly, sometimes with distinctive "crowfeet" indicating the spreading of flow. The phenomenon is reaffirmed by textual evidence, including land-grant case testimony. Nathan Coombs, in 1861, testified about how streams on the west side of the valley (in the Caymus grant) spread into wetlands:

> *Do not the streams which come from the mountain on that side flow into Napa River in the Wet season? And if not, what becomes of them?*
>
> I think the streams spread and sink on the plain which makes the wet and marshy ground I have just described.[7]

Even in the 1930s—after nearly a century of drainage efforts—Carpenter and Cosby observed that many creeks still entered the valley in "shallow channels that are lost on the lower fan slopes."[8] Spreading streams helped retain water in wetlands and groundwater, reducing flood energy and slowly percolating flow to the river through the dry season, rather than rapidly delivering it to the Bay, as occurs today.

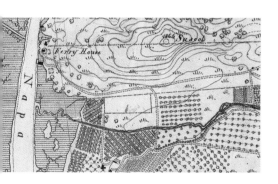

Early creek modification. For some creeks, the earliest maps available show well-defined channels, but other documents indicate that these channels had already been modified. For example, Suscol Creek is shown with a continuous channel into tidal Napa River in the 1858 T-sheet, but Menefee described how the stream had been extended several years earlier: "Soscol Creek, which is now confined within artificial bounds and empties into the river, in 1852 spread over a wide area, converting it into a morass. This is now reclaimed and constitutes the richest portion of the Soscol orchards."[9] *Kerr 1858, courtesy of NOAA.*

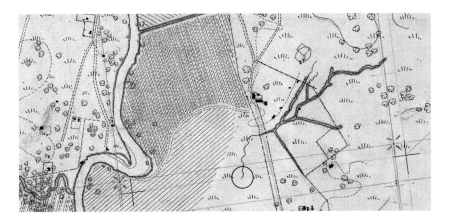

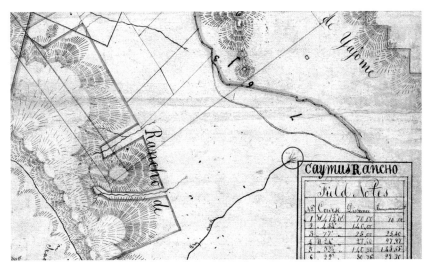

Spreading streams in historical maps. Mid-19th-century surveyors depicted a number of creeks terminating on the valley floor. For example, Kerr (1858; top) showed Tulucay Creek becoming a smaller and shallower channel before ending a half mile from the nearest tidal channel; Dewoody (circa 1860; middle) illustrated the spreading of Dry Creek and its neighbors to the north; and the original *diseño* for the Carne Humana land grant (USDC circa 1840a; bottom) showed two major creeks flowing across the valley between Bale's house and the Calistoga hot springs, each approaching but not connecting to the Napa River. *Top: courtesy of NOAA. Middle: courtesy of The Bancroft Library, UC Berkeley. Bottom: courtesy of The Bancroft Library, UC Berkeley.*

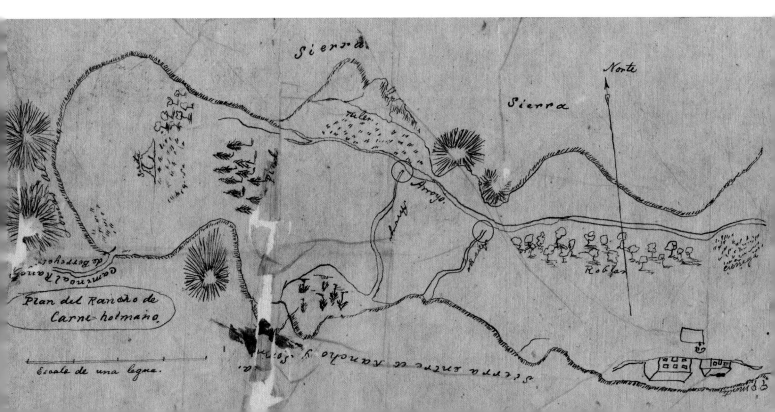

THE HIGH WATER TABLE PROBLEM

Poor drainage and a fluctuating water table result in much damage to trees and vines throughout the valley.

—SOIL SCIENTISTS
E. J. CARPENTER AND
STANLEY W. COSBY, 1938

The diffuse nature of the Napa Valley's stream network was one of the main limitations to its use for agriculture. The system of spreading streams, valley wetlands, broad floodplains, and natural levees tended to both retain water on the surface of the land and store it as groundwater below. The surface soils of the valley filled like a sponge in the winter, gradually drying through the summer and fall. It was the high winter water table, even more than seasonal flooding, that necessitated the transformation of the stream network and the draining of the valley.

To survive the long dry season without irrigation, orchards and vineyards—like oaks—must establish deep root systems to reach moist soils. But across much of the early Napa Valley, the water table rose within a few feet of the ground during the wet season. While valley oaks and other native species could handle this dynamic setting, the extended soil saturation restricted root development in crops: deep roots would drown.[10] Unlike native flora adapted to these seasonal fluctuations, agricultural plantings suffered from the extreme natural variation in groundwater levels.

Soil scientists Carpenter and Cosby recommended a widespread increase in drainage to eliminate the high groundwater problem, particularly in the lowlands along the Napa River: "The soils in the central part of Napa Valley, almost without exception, would be benefited by artificial drainage."[11] Saturated lowland soils could be dewatered by creating ditches or "drains" that would convey water from the valley floor directly into the river. The success of this aggressive expansion of drainage—much of which took place in the 19th century—has expanded agricultural potential but also reduced groundwater recharge and contributed to flood peaks downstream. More recently, the construction of urban storm drains has also accelerated the conveyance of water to the river.

The Napa Valley remains blessed with a high natural recharge capacity. During most years, the valley's surrounding watershed receives sufficient rainfall to refill the Napa Valley aquifer, recovering from the annual dry season decline. In the Napa Valley, unlike many other parts of California, 20th-century agriculture did not cause the widespread decline of groundwater levels.[12] However, increases in urban use as well as in the extent (and potentially the intensity) of agricultural land use, when combined with a series of low rainfall years, are likely to produce water shortages in the future.[13] Recognizing the value of near-surface groundwater to streams and the local water supply, farmers are increasingly employing new drainage techniques to maximize percolation. As water security and reuse become higher priorities, reconsidering the structure and function of the drainage system could produce new strategies for sustaining both the agricultural and the ecological functions of the valley.[14]

Agricultural bounty, a century ago. These postcards, produced around 1910, celebrate the orchards that thrived without irrigation in the naturally moist Napa Valley soils.[15]

PLUMBING THE VALLEY

In many places deep open drains leading into Napa River would materially improve the drainage conditions.

—E. J. CARPENTER AND
STANLEY W. COSBY, 1938

To create today's drainage network, 19th-century farmers dug new channels to connect the many discontinuous, spreading streams more directly to the river. Because it took place so early, this transformation has been largely forgotten; many artificial channels have now existed for more than a century. During the same time, about one-third of the valley's historical channels have been filled.

As a result of these efforts, the contemporary creek network reflects a complicated combination of natural and constructed channels. Many creeks follow sinuous, natural courses for some distance from their canyon mouths before abruptly shifting to artificial, straight-line extensions. A number of creeks that ran parallel to the Napa River before finding a low point in its natural levee now flow directly into the river. Entirely new, rectilinear creek networks have been created to remove water from valley areas that formerly had no natural drainage.

Changes in the hydrograph. A hydrograph is a graph of water flow past a given point on a river over time; an annual hydrograph shows how flow varies throughout the year. As part of an analysis of water-sediment dynamics in the Napa River watershed, information from the historical ecology study was used to explore how the historical (dashed black line) and modern (blue line) annual hydrographs might differ, using a hydrologic model of the watershed.[16] The results of the model, while not precise, suggest that the way water moves through the watershed has changed dramatically, resulting in much larger peak flows today, as well as reduced dry season flows.

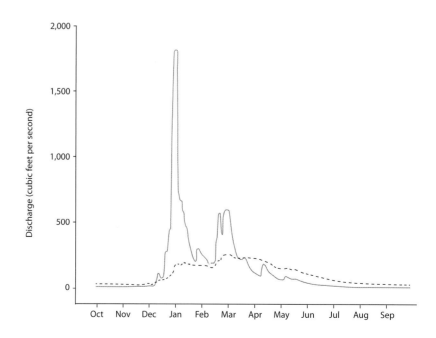

Expansion of the drainage network. Comparison of historical and contemporary[17] mapping of Napa Valley streams shows that the total length of above-ground channel (including the Napa River) on the Napa Valley floor has increased by about 25%. This overall increase has occurred despite the elimination of about one-third of the historical drainage network (i.e., creeks that have been completely filled or buried). As a result, of the contemporary channels draining the valley, about half are artificially constructed, and about half follow their historical alignment. As shown in the insets, in some places extensive drainage networks are almost entirely artificial.

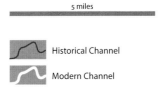

RIPARIAN LOSS AND RECOVERY

The corridors of riparian trees alongside creeks exhibit several different trajectories. As would be expected, there has been significant clearing of these narrow forests. The filling or covering of historical creeks has resulted in loss of the associated riparian habitat. Cities and agriculture have encroached on existing creek corridors as well. The narrowing and fragmentation of Napa Valley's riparian forest is reflected in the decline of some riparian birds, such as the yellow-breasted chat, and the extirpation of others, like the yellow-billed cuckoo.[18]

On many creeks, however, riparian forest is now more prominent and continuous than it was in the earliest aerial photographs, about 70 years ago. Streams such as Ritchey Creek east of Calistoga and Sulphur Creek in St. Helena exhibited large gaps in riparian vegetation in the 1940s, when orchards closely impinged upon the channel. Today these gaps have filled as vegetation has regrown.[19] Riparian trees even reoccupied Napa Creek, which ran underground through downtown Napa for a half century (see pages 64–65). While much riparian forest has been lost along the Napa River during the second half of the 20th century, many of its tributaries show an increase in vegetation during the same time.

Another interesting landscape response has occurred on some of the new channels created to drain the valley floor. These artificial creeks were conspicuous in the 1940s for the lack of riparian vegetation. But in the intervening decades, native trees have colonized some of the constructed channels, creating new riparian habitat. The spontaneous establishment of riparian forest on previously denuded (or even completely artificial) channels demonstrates the resilience of the native landscape. Where we have accidentally or intentionally provided the small amount of space these habitats need, they have often responded dramatically.

[I] was awakened soon after daylight by music new to my ears, but so delightful and sweet, the impress on my memory has never been dimmed. It was the singing of hundreds of various kinds of wild birds, living and nesting in the trees and brush bordering the stream flowing back of the hotel.

—FRANK LEACH, RECOLLECTING NAPA CREEK IN THE SPRING OF 1857[20]

Riparian trees along an artificial channel. This line of trees along Sulphur Creek, just upstream of Main Street in St. Helena, has developed since the stream was moved from its historical location to a new ditch in the 1940s (see page 63). *Photograph March 23, 2003.*

Changes on Ritchey Creek, 1942–2009.
A reach on lower Ritchey Creek appears to have been significantly cleared by 1942 **a**, presumably for orchard development. In a 2009 image, crops encroach less closely on the channel, and the riparian forest has expanded **b**. Paralleling the river, a now-disconnected portion of the creek **c** has more riparian forest than it did in the 1940s.

UNDERCUT TREES AND CHANNEL INCISION

One of the concerns about Napa Valley's creeks is an observed trend toward downcutting—a process called channel incision. As in other parts of the American West, many creeks appear to occupy extremely deep channels. This pattern is commonly seen under bridges, where piers have been exposed well below their intended bases, often weakening the structural integrity of the road itself (see page 96). In some places, incision has left dramatic evidence in the form of mature trees that have been undercut by the erosion of the channel bed and bank, leaving their roots exposed high above the stream. Examining the age of the tree, the amount of undercutting, and even the tree's growth in response to the channel changes can help estimate the timing and amount of incision.[21] On Carneros Creek these clues suggest downcutting of up to 6 feet since the early 19th century in certain reaches.[22]

Incision is driven by a change or imbalance between sediment supply and stream flow.[23] The process can work both downstream and upstream along channel networks. For example, on-stream reservoirs retain sediment, causing them to release "hungry" water, which, without the balance of sediment, has more erosive energy. The greater degree of connectivity in the contemporary creek network concentrates stream flow and increases channel erosion. Land-use activities that cause a lowering of the main stem channel bed (in this case, the Napa River) can also work backwards up the drainage networks, causing incision to "unzip" the tributaries. The connection of

shallow, formerly discontinuous tributaries to the deeper main stem Napa River also has presumably caused their bed levels to lower.[24]

Channel incision results in higher and often steeper stream banks. If sufficient downcutting occurs, stream banks can fail, resulting in channel widening.[25] As banks have deepened and widened along local creeks, many riparian trees have become severely undercut, threatening both the persistence of a mature riparian corridor and the stability of stream banks and adjacent property.[26]

Management at the watershed scale to reduce the erosive potential of creek flows can reduce incision, benefiting both creeks and their neighboring residents. At the local scale, strategies to allow targeted bank erosion can reduce the risk of unpredictable stream bank failure while recovering riparian functions.[27]

Erosion of Napa Valley stream banks. Bank retreat, often associated with incising streams, can threaten adjacent property and structures (opposite page, left). Undercut California bay laurel trees (opposite page, right) can adapt to changing bed and bank structures by rotating their root balls, maintaining equilibrium by sending roots up the bank. This rotation takes several decades and is likely a remnant of 19th- or early-20th-century downcutting.[28] Other riparian trees, such as valley oak (left), do not exhibit such morphological adjustment and must survive with their roots exposed midair as banks retreat. *Opposite page: photographs by Julie Beagle, June 19, 2009 (left) and October 25, 2008 (right). Left: photograph by Lisa Micheli, June 3, 2009, courtesy of the Tessera Sciences 2009 Napa River Rutherford Reach Annual Stream Monitoring and Maintenance Survey.*

BRAIDED CHANNELS

While most Napa Valley creeks exhibited a narrow, tree-lined morphology, some demonstrated a very different character. Early maps and descriptions reveal several braided streams—wide, gravelly channels with multiple flow paths and few trees. In response to a large sediment supply and high energy, these alluvial fan streams were especially dynamic, producing broad active channel beds composed of unvegetated, frequently shifting gravel bars and riparian scrub vegetation.

Because braided channels are relatively uncommon in the Napa Valley, the few examples have sometimes been interpreted as artificial byproducts of previous land use. But they are confirmed historically and correspond to appropriate geology and sediment supply. Sulphur and Conn creeks—the two largest braided channels in the Napa Valley—are each associated with the naturally erosive Franciscan bedrock formation. Braided channels were often mapped as "Riverwash" in early soil surveys, described as "water-worn cobbles, gravel, and sand…[that] support little or no vegetation, although in some places, in areas protected somewhat from the full sweep of flood waters, willow, weeds, and grasses have become established."[29] Early maps, descriptions, and aerial photography reiterate these patterns.[30] Vegetation was limited and included few mature trees; frequent channel movement scoured out saplings and continually restarted successional patterns.

While braided streams are a natural component of the California landscape, they are often unappreciated, perhaps because of their

Braided channel reach on Sulphur Creek. Flows naturally shift between multiple channels as floods rearrange the pattern of bars, channels, and vegetation. Gravel was mined from Sulphur Creek for decades until 1999. Ironically, these dynamic, unstable channels provided the structural materials for buildings and roads throughout the valley. *Photograph looking upstream from the Crane Street bridge, March 23, 2003.*

somewhat "scruffy" appearance.[30] However, braided channels can support an array of bird species and provide spawning habitat for steelhead. They represent a distinctive Napa Valley stream type—a highly organized, energy-dissipating system associated with frequent natural disturbance.[31]

Historical gravel mining can be another indicator of braided channel conditions. Both Sulphur and Conn creeks supported long-term gravel mining operations that relied upon annual deposition for supply. Sulphur Creek was surface-mined for many decades without creating a giant pit, and a local resident relied on the annual gravel supply to Conn Creek until the construction of Conn Dam (protesting that the dam ended her business).[32] Interestingly, in recent years commercial gravel mining on Sulphur Creek has ceased and the channel has begun to aggrade, or build up, providing an opportunity to observe a braided stream with relatively natural channel form and sediment supply. As an alluvial fan channel with a high natural sediment load, the creek will tend to fill its channel and, at some point, initiate a new course across its fan.

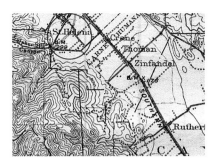

Sulphur Creek, circa 1899. The first USGS map of Napa Valley, surveyed during 1896 and 1899,[33] distinguishes Sulphur Creek as a braided channel by a dotted pattern indicating sand or gravel, rather than the standard blue line used to represent other creeks. *Courtesy of the Earth Sciences and Map Library, UC Berkeley.*

Conversion of a braided channel to a single thread channel. This photo sequence shows how a segment of Sulphur Creek was removed upstream of Main Street in St. Helena. The braided 1942 channel extent is shown with a yellow outline in both images, with the main channel indicated by a dashed blue line. The new, straight channel (shown with a solid blue line) was created by connecting Sulphur to an existing ditch, the tributary Spring Creek. With the additional flow, this channel has developed a narrow, tall canopy (see page 58) that contrasts with the dispersed, low riparian scrub vegetation that characterized the earlier channel.

SUMMER WATER

Water has always been scarce during the hot Napa Valley summer. Even some of the largest valley creeks naturally went dry, or nearly dry, in their lower reaches, sinking into their porous alluvial fans.[34] Captain W. F. Wallace noted, "Dry creek [one of the valley's largest watersheds] rises in the mountains west of Yountville, and as its name indicates, is almost dry in the summer."[35] Insurance maps identifying potential water sources for firefighting consistently labeled lower Napa Creek, fed by moist redwood canyons, as "dry in summer."[36] Bartlett Vines testified in 1861 about Conn Creek, another major tributary: "In the summer season, or dry weather, it does not run out into Napa Valley."[37]

Yet intermittent stream channels can maintain pockets or pools of water, as the Wallace quote above suggests ("almost dry"). Scattered persistent pools, connected by subsurface flow, can provide critical habitat for native fishes to survive the summer. In fact, the first recorded description of a Napa Valley creek noted that Carneros Creek, in midsummer 1823 near Old Sonoma Road, had some flow, with pools likely suitable for steelhead and salmon:

> Presently we arrived at an arroyo which was said to be the entrance to Napa. This measures very little flowing current, of not much abundance are their waters, but we observed by measuring some places there are small permanent ponds with clear, sweet, abundant and pleasing water, sufficient to water some cattle.[38]

Residents living along the lower reaches of Carneros Creek over the past half century describe similarly intermittent conditions, noting that the stream "didn't run year-round" and that "every year it dries up."[39]

Changes in flow on naturally intermittent streams are difficult to document, but there is some anecdotal evidence that local streams have become even drier. For example, interviews with a number of residents along Carneros Creek independently described a diminution in the extent of pools and seasonal flow, reporting that "it ran more" and "used to visibly run… enough to get over the rocks."[40]

Long-term well records do not indicate major declines in groundwater levels in most parts of the valley.[41] However, the more rapid decline of near-surface water each spring caused by expanded drainage and stream/river incision has the potential to reduce the amount of dry season "base" flow in Napa Valley's creeks.[42] Pumping from streams and the groundwater aquifer exacerbates these underlying changes in the valley water cycle. The natural sponge effect of the floor of Napa Valley has been reduced: the sponge doesn't wet as fully as it used to, and it dries out more quickly.

"Dry in Summer." (opposite page) Sanborn insurance maps evaluated Napa Creek as a potential water source for fighting fires downtown. This map shows the Napa Opera House and other downtown buildings, several of which were actually built across the creek. *Sanborn Perris Map Company 1886, courtesy of the California State University Northridge Map Library.*

Bale Mill. (below) Perennial flow on Bale Creek powered the famous grist mill. Bale and neighboring Ritchey creeks appear to have been among the valley's rare year-round creeks. *Photograph circa 1934, courtesy of the Library of Congress.*

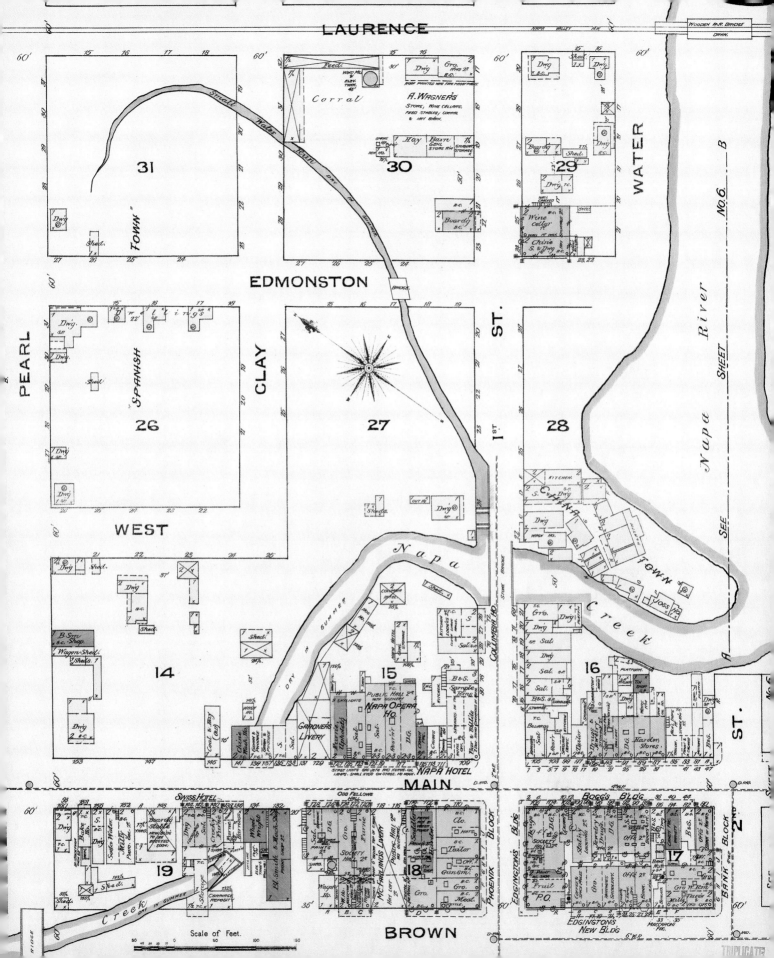

"April Showers, Napa Valley." California landscape painter Jules Tavernier shows a characteristic early Napa Valley scene of seasonal wetlands surrounded by wildflower meadows, circa 1880. *Courtesy of the California Historical Society.*

4 • VALLEY WETLANDS

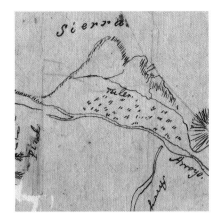

Valley wetlands in historical maps. *Top and middle: courtesy of The Bancroft Library, UC Berkeley. Bottom: Kerr 1858, courtesy of NOAA.*

State as near as you can the quality of the land within that part of the survey.

There are different qualities—good, dry land, wet marshy land and hills. The dry land is good for cultivation—the marshy land can be made tillable land by drainage[—]with present condition it cannot be cultivated.

—TESTIMONY BY BARTLETT VINES IN THE CAYMUS LAND-GRANT CASE, 1861

The stories of oaks and creeks suggest some of the seasonal drama of the historical Napa Valley landscape. In the rainy season, creeks spread out and merged with river overflow to temporarily flood the bottomlands. Rising groundwater approached the land surface. Wetlands filled from rainfall and runoff.

By summer, the landscape had transformed. Floodwaters sank into the ground. Most streams ran dry as the water table fell. Valley oaks, well spaced to avoid water competition, dominated the landscape and reflected the scarcity of surface water. Napa's streets, famously muddy in the winter, turned to dust.

As the valley dried, the few perennial (year-round) freshwater marshes became oases, lush watering holes of ecological activity within the mostly dry surrounding savannas. Some fishes and amphibians would spend nearly their entire lives in these marshes, while juvenile steelhead, migratory waterfowl, and even grizzly bears would use them at certain times of year. The loss of these habitats has driven species such as thicktail chub, a fish once common in these kinds of wetlands throughout the state, to extinction. Many wetland-associated species are now rare in the valley, if not completely extirpated.

In the Napa Valley's Mediterranean climate (with no snow pack and an annual five-month drought), wetlands were limited to specific places in the landscape, where topography and groundwater combined to provide extended access to surface or near-surface water. Seasonally wet meadows occupied flat or gently sloping areas with heavy, water-retaining clay soils. Permanent freshwater marshes formed where peculiarities of local topography created natural depressions with access to groundwater. Relatively rare wetland types, such as vernal pools and alkali meadows, occurred in seasonally flooded areas with unusual soil characteristics. The river itself spread broadly into riverine wetlands in several places; these are described in the Napa River chapter. (In addition, tidal wetlands associated with the Bay were found in the southern end of the valley and are discussed in the Tidal Marshlands chapter.)

As the wettest parts of the valley's natural sponge, wetlands captured surface water in the wet season and gradually released it downstream during summer, helping prolong stream flow. Valley wetlands also trapped fine sediment, which would settle out as floodwaters slowed in low areas.[1]

The expansion of valley drainage has been extremely effective at eliminating the valley's wetlands. However, basic topographic rules are still relevant today—low areas are still low and prone to flooding. Groundwater levels still approach the surface in places;[2] clay soils retain their water-holding character. The former distribution of wetlands may suggest opportunities to restore Napa Valley's wildlife habitats.

This chapter investigates the valley's wetlands in five parts. We first examine the characteristics of freshwater marshes, wet meadows, vernal pool complexes, and alkali meadows, and then look at the overall changes to the wetland landscape.

Just about where the town of Rutherford stands, there used to be large patches of wild blackberries, covering many acres. Here in berry time the bears would gather.

—IGNACIO VALLEJO[3]

What is the nature of the Country between the Spring and the River as to Dryness?

It is generally marshy low ground where the stream sinks.

How is the road constructed through this marsh?

There are ditches on each side and the road is banked up above the natural surface.

—C. C. TRACY, 1861

A contemporary marsh. This wetland—the largest remaining natural freshwater marsh in the Napa Valley—receives flow from Old Faithful Geyser, just west of Calistoga. *Photograph by Susan Schwartzenberg, November 22, 2008.*

1846 April the 1st.... Left Mr. Younts and proceeded down Nappa vally [sic] through several sloughs and mud holes.

—JAMES CLYMAN, 1846[5]

I can not place the swamp indicated on said exhibit, though there are swamps in the valley.

—BARTLETT VINES, TESTIFYING ABOUT A RANCHO MAP IN 1861

VALLEY FRESHWATER MARSH

Freshwater marshes appear in some of the earliest documents of the valley. Mexican *diseños* labeled these wetlands as *ciénaga*, *pantano*, or *tular*—all referring to a type of marsh or swamp—and the features are discussed in accompanying land-grant testimony. American surveyors noted entering and leaving "swamps," "tule," and "low boggy land"—terms indicative of perennial wetland conditions. Historical USGS maps used the conventional symbol of broken horizontal lines with tufts (indicating vegetation in water[4]) to show perennial marshes. The mottled, moist soil signature of many of these features was still evident at the time of early aerial photography, helping to confirm their spatial extent.

In the early 1800s, there were at least eight freshwater marshes larger than 10 acres on the valley floor. Four of these—all located in the broadest part of the valley, between Oak Knoll and St. Helena—were extremely large systems, greater than 100 acres in size. Historical sources document about 800 acres of valley freshwater marsh in total, an amount which would fluctuate to some extent during drier and wetter years and decades. We also identified nine perennial ponds, ranging from less than 1 acre to 5 acres. These are probably conservative estimates, given the likely presence of many small undocumented wetlands.

While the historical extent of valley freshwater marsh was great compared to today, these features were still relatively rare. They occurred in distinct landscape positions—between coalescing alluvial fans (e.g., Bale Slough), behind knolls (e.g., east of Dunaweal Lane near Calistoga), at natural springs (e.g., Calistoga)—that supplied year-round water.

Cinnamon teal in a California wetland. Waterfowl like the cinnamon teal and ruddy duck nested in the valley's freshwater marshes, but are now largely limited to marginal sites at reservoirs and wastewater treatment plants. The now-rare tricolored blackbird likely nested in the cattails and tules of the valley's marshes, while red-legged frogs probably bred in the open water. *From Grinnell et al. 1918, courtesy of UC Press.*

VALLEY WETLANDS • 71

A natural knoll dam. Just east of Dunaweal Lane, two of Napa Valley's knolls nearly touched, blocking surface and subsurface flow to create a 20- to 30-acre freshwater marsh. A portion of the wetland was identified as late as 1943 in the USGS Calistoga quadrangle (bottom right). In 1942 aerial imagery (top left), wetland patterns are apparent near the knolls **a** and suggestive of former wetlands over a broader area that had no orchards **b**. By 2005 (bottom left), the wettest portion had been developed into a reservoir **c**; some residual wetland character may still persist at other sites **d**. *Bottom right: USGS 1943, courtesy of the Earth Sciences and Map Library, UC Berkeley.*

WET MEADOWS

A marvelous wealth of the yellow monkey flower rioted in the lowlands.

—WILLIS JEPSON, 1912A

The most common wetlands in Napa Valley were wet meadows: temporarily flooded lowlands dominated by sedges, rushes, grasses, and wildflowers (also called wet prairie, moist grasslands, temporarily flooded grasslands, or seasonal wetlands).[6] Wet meadows were typically positioned in broad, nearly flat or basin-like areas, occupying fine-grained, clay-rich sediments with poor natural drainage. Seasonal ponds formed for days or weeks associated with rainfall events, providing foraging habitat for migrating waterfowl and juvenile fish. Trees were generally excluded by the limited drainage.

The characteristics of wet meadows usually result in a distinctive land-use history. Orchards, vineyards, and other high-value, deep-rooted crops tend to be excluded to some degree from these areas, or require extensive drainage, plowing, and/or irrigation to be successful. As a result, wet meadows often remain under grazing or grain production for a longer period of time than the adjacent, better-drained lands.

In the Napa Valley, as in other parts of California, the single best indicator of former wet meadows is the historical and modern soil surveys of the USDA. Soil survey teams have passed through the Napa Valley three times in the past century: Holmes and Nelson (surveyed in 1914, published in 1917), Carpenter and Cosby (1933/1938), and Lambert and Kashiwagi (1965–73/1978). Each team mapped clay-rich soils with drainage problems and agricultural limitations, and described these characteristics in an accompanying report. We translated these indicators into a map of historical wet meadows, selecting thirteen soil types mapped by Carpenter and Cosby and two soil types mapped by Lambert and Kashiwagi.[7] Typical descriptions include "soil becomes boggy after heavy rains," "because of its heavy texture and poor drainage, its value for agriculture is low," and "groundwater level stands near the surface during the wet season."[8]

Botanical accounts support our interpretation of these areas. Jepson described plants common to moist meadows, such as the yellow monkey flower and the large-flowered star tulip—the latter found "in wet lands at Calistoga." He collected other species in the "wet meadows near Rutherford," likely corresponding to the Bale Slough area.[9] Showy Indian clover, typically found in low, wet swales, was found near Napa Junction in 1891.[10] USDA soil surveys identified wet meadow soils at each of these localities.

VALLEY WETLANDS • 73

Wet meadows and farming. Soils such as the Zamora silty clay and adobe clay (Zs and Za, above), which were described by Carpenter and Cosby as "in need of drainage,"[11] remained in hay or grain production in 1942 (top left), after surrounding areas had already been converted to orchards ⓑ. These sources of information helped us to estimate the extent of wet meadows in the early 1800s (bottom left).

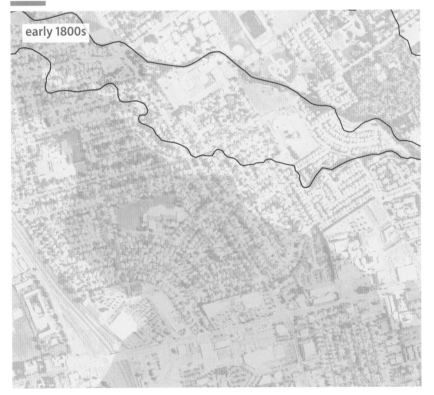

Wet Meadow

Oak Savanna

VERNAL POOL COMPLEXES

Popcorn flower. (above) A specimen collected by Jepson, in "Napa Valley near Calistoga" on May 2, 1897. *Courtesy of the Consortium of California Herbaria.*

Douglas' meadowfoam. (below) At the Soscol Creek vernal pools. *Photograph by Jake Ruygt, April 2010.*

As with all aspects of the native landscape, it is difficult to document the full historical complexity of the Napa Valley's wetlands. Clues are provided by the specimens collected by local botanists from the 19th century through recent decades.

Vernal pools and swales are seasonally flooded depressions that support a number of unique plant species. Pools, swales, and intervening hummocks often occur in larger complexes with a distinctive undulating topography. Compact subsoils hinder drainage. These short-lived seasonal wetlands are famous for the dramatic rings of color created by the sequential flowering of different species as pools dry from the edge inward. Waterfowl such as mallards and cinnamon teal often nest near larger pools.

Because most plant specimens were not recorded with precise spatial data, only a few places can be specifically mapped as historical vernal pool complexes. However, the distribution of plants associated with vernal pools suggests that these features occurred much more widely through the valley. *Downingia* species, including the maroonspot, toothed, and dwarf calico flowers, were found near Yountville, Oak Knoll, Napa, and in the wet meadows south of Napa. Goldfields were recorded in the vicinity of Napa, St. Helena, and Calistoga. Between 1892 and 2000, *Plagiobothrys* species—including the Calistoga, rusty, slender, and bracted popcorn flowers—were collected in the vicinity of Suscol, Napa Junction, Rutherford, St. Helena, and Calistoga. (In fact, many of California's most illustrious botanists, including Jepson, Alice Eastwood, Milo Baker, Herbert Mason, and Peter Raven, came to the Calistoga wetlands to see the diverse flora.)[12] Vernal pool species were still recorded at a number of sites into the 1980s, but most of these populations have now been extirpated.[13]

Vernal pool/swale landscape near Oak Knoll. Vernal pool complexes appear to have been extensive in the lowlands east of Napa River between Yountville and Oak Knoll Avenue. The area shown here was described in 1917 as having "slight hog-wallow topography"—a common description of vernal pool complexes.[14] Early aerial photography (above) shows the expression of this distinctive environment at the landscape scale: darker interconnected pools and swales are separated by lighter-colored hummocks, across an area of over 200 acres. By 2009 (left) the area had been transformed. (The later image is shown at smaller scale.)

ALKALI MEADOWS

Alkali meadows are relatively rare plant communities associated with high concentrations of soluble salt resulting from the annual evaporation of shallow overflows, especially those with high mineral content.[15] They have soil and drainage characteristics similar to those of wet meadows but support a distinctive flora tolerant of high salt concentrations, including a number of now-rare plant species. In the Calistoga area, botanists have collected plants from alkali meadows for over a century, describing specimens from the "alkaline flats" and "on alkaline flat, around all the alkaline springs."[16]

A number of species associated with alkali meadows have also been recorded just above the historical Bay margin in the lower Napa Valley, suggesting a zone of salt effects associated with occasional tidal influence

and/or seasonal flooding from rainfall. These species include sack clover, alkali milkvetch, and San Joaquin spearscale. The latter species has been found in recent decades near the tidal marsh margin both south of Napa and west of Vallejo.[17] These botanical data begin to illustrate a high tidal marsh ecotone with alkali meadow characteristics, as has been observed at Bay margins elsewhere.[18]

These rare surviving or now-extirpated plants hint at the botanical richness of the valley's former wetlands. The association of these habitats with relatively persistent physical characteristics (e.g., claypan, salt-affected soils) suggests limited but specific areas of potential recovery, depending on the intensity of historical land use.

Alkali meadows at Calistoga. In the background of this postcard view of the Calistoga hot springs (above), patterns of dry and wet vegetation, ponds ⓐ, and bare playas or scalds ⓑ can be seen. The mineral content of the waters has led to distinct soil characteristics and associated plants. Some of the same area can be seen in the March 20, 1907, Turrill and Miller photograph (left), as well as steam spouts associated with hot springs. Salt-tolerant plants such as saltgrass are still found in fields and yards in the Calistoga area. *Postcard courtesy of Todd Schulman. Photograph courtesy of the California Historical Society.*

WETLAND FRAGMENTATION AND RESILIENCE

The wetlands explored in the previous pages are among the most severely impacted components of the Napa Valley landscape. To evaluate changes since the early 1800s, we used recent mapping by the Bay Area Aquatic Resources Inventory, which has detailed the distribution of wetlands within the contemporary landscape.[19] While historically there were about 13,000 acres of valley freshwater marsh, wet meadow, alkali meadow, and vernal pool complex, about 100 acres exist today—declines of 99%–100% for each wetland type. Some of the water that supported these wetlands is now directed to a new kind of feature that dominates the wetland landscape—artificial bodies of open water constructed for water storage.

By comparing past and present wetland mapping, we can also investigate the origin of today's wetlands. Some features may be remnants of historical wetlands, maintained largely by natural processes, while others may be associated with ditches, reservoirs, and other constructed features. In the Napa Valley, we found that 70% of the contemporary freshwater marsh and wet meadow is in fact a remnant of former wetlands.[20] Furthermore, over 30% of today's freshwater marsh and wet meadow derives from the valley's historical freshwater marshes, despite the fact that these marshes occupied less than 2% of the valley. Most of the largest contemporary wetlands are found in these spots—small, seasonally wet meadows persisting within the footprint of much larger former tule marshes. The few remaining alkali wetlands also are located within historical alkali meadows, because of the persistence of soil salinity. Interestingly, a number of the larger remnant wetlands are bordered by undeveloped lands, presumably because poor drainage characteristics limit their value for agriculture and other intensive development.

The correspondence between the past and present landscape suggests that despite the extensive modifications to drainage patterns, the ecological fundamentals are still intact. The wettest parts of the valley are still wet—the landscape still guides water to the same places, and they tend to retain water. As efforts are made to recover some of the valley's missing wetland functions, these persistent features offer significant opportunity. The historical landscape provides a template for identifying possible restoration nodes, revealing the places that have naturally sustained wetland mosaics over the long term and will be likely to do so in the future.

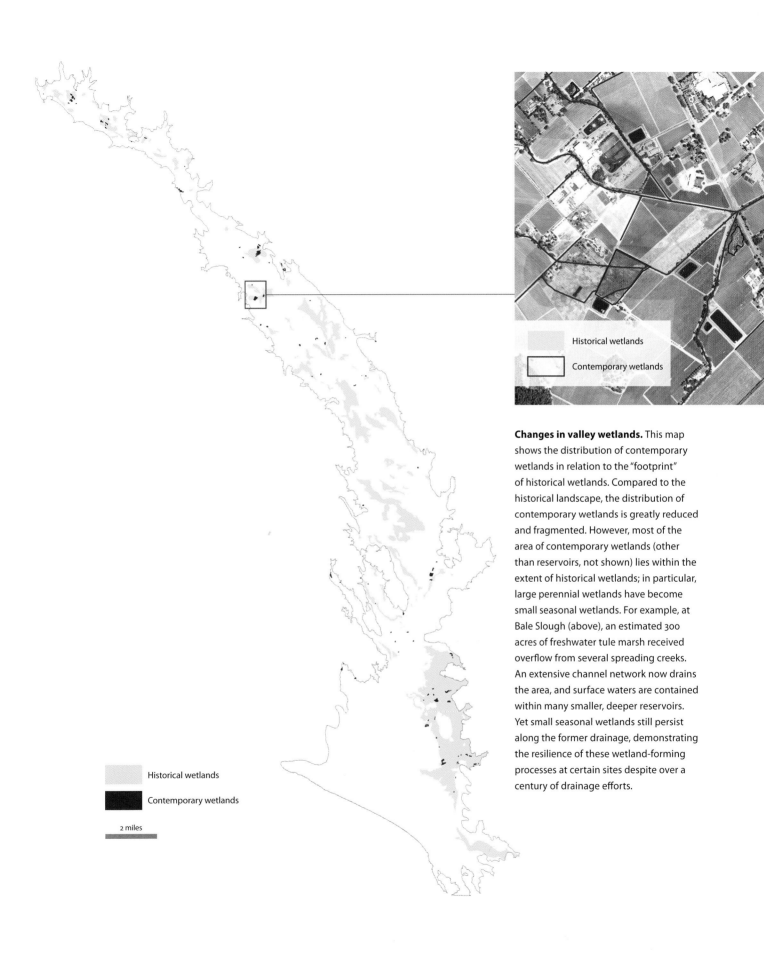

Changes in valley wetlands. This map shows the distribution of contemporary wetlands in relation to the "footprint" of historical wetlands. Compared to the historical landscape, the distribution of contemporary wetlands is greatly reduced and fragmented. However, most of the area of contemporary wetlands (other than reservoirs, not shown) lies within the extent of historical wetlands; in particular, large perennial wetlands have become small seasonal wetlands. For example, at Bale Slough (above), an estimated 300 acres of freshwater tule marsh received overflow from several spreading creeks. An extensive channel network now drains the area, and surface waters are contained within many smaller, deeper reservoirs. Yet small seasonal wetlands still persist along the former drainage, demonstrating the resilience of these wetland-forming processes at certain sites despite over a century of drainage efforts.

"Napa Valley and River", 1885, by Manuel Valencia. In this painting, the Napa River is shown with a narrow riparian corridor, which may have been influenced by early clearing. The river is relatively shallow and well connected to its floodplain. Based on the broad, open channel, the view is probably of the lower fluvial reaches. *Courtesy of the Collection of the Saint Mary's College Museum of Art, gift of James J. Coyle and William T. Martinelli.*

5 • NAPA RIVER

In the midst of the valley winds a small stream…its course marked by the graceful willows that grow along its margin.

—U.S. COMMISSIONER JOHN BARTLETT, 1852

The headwaters of the Napa River can be found in the deep canyons of Mt. St. Helena, on the northern margin of Napa Valley. Seeps and springs trickle into small creeks that emerge onto the valley floor, initiating the complex, interacting system of vegetation, sediment, and water we call the Napa River. Today, the river mostly follows a single deep channel, with long stretches of flat, open water and a narrow riparian corridor. It is easy to assume that the tree-lined, still-sinuous river retains its original character. But historical records show that the Napa River looked and functioned very differently in the recent past. When compared to former conditions, the present river turns out to be simplified and constrained, retaining only remnants of its previous character.

In the early 19th century, the Napa River grew and changed as it headed downvalley. It received mineral-rich overflows from the hot springs at Calistoga, met perennial tributaries at Mill and Ritchey creeks, and gained copious sand and gravel inputs from streams in its middle reaches. Alluvial fans and bedrock knolls deflected the river's course from side to side. As the river approached San Pablo Bay, it widened into a broad, slow-moving estuarine channel, spreading out to nourish the vast tidal marshlands.

Upper Napa River, June 2, 1907. A Turrill and Miller photograph shows riparian trees in leaf and substantial water in early summer. The caption "Upper Porstion *[sic]* Napa River" is not geographically precise (nor spelled correctly) but refers to the river upstream of St. Helena, where the channel was narrower and bordered by dense riparian forest. *Courtesy of the Society of California Pioneers, catalog no. C013708.*

The Napa River responded to the width and slope of the valley floor, supporting broad riparian forests in some reaches and floodplain wetlands in others. Over 30 miles of side channels branched out across the valley bottom, providing forage and refuge for young steelhead. Native minnows lived in warmer water and were a major part of the native diet; steelhead and likely chinook salmon grew in the deeper pools. Multicolored wood ducks nested above slow-moving waters; wild grapes climbed oaks, bays, and willows. Every summer for thousands of years, people fished and swam in the river's cold pools and reclined on its gravel beaches.

The Napa River remains the valley's central geographic feature. It is the spine of the valley, the link connecting hills and tributaries to the Bay. The river's tidal landings enabled prosperous commerce in the days before the railroad. Riverine flooding created the fertile land supporting today's agriculture. During heavy rains, the river provides the only conduit for transporting high flows safely through the valley and into the Bay. Native fish and wildlife still rely on the Napa River; it is considered the most important watershed for steelhead recovery in the entire Bay Area.[1] And despite its constriction by adjacent levees and land use, the river continues to be dynamic, eroding and flooding unpredictably.

Like many California valleys, the Napa Valley is naturally flood-prone, with at least 20 significant floods since 1850.[2] Floods in 1995 and 1997 each caused over $100 million of damage. Since at least the 1870s, when a 3-mile levee was built in the Yountville reach,[3] earthen walls have been constructed along the river to prevent high flows from occupying the floodplain. However, during the same period, the increasing density and connectivity of the drainage network, and the removal of floodplain capacity through levee construction, have directed more water more quickly into the main channel, at least partially counteracting these efforts. While flood extent has been reduced, flooding has not gone away—in fact, downstream flooding may have increased.[4] And the river's banks are now subject to greater erosion by the confined flows.

In the interest of flood control, most of the river was nearly converted to a heavily engineered channel, as the U.S. Army Corps of Engineers proposed an extensive project to widen the river in the decades following the 1955 flood.[5] Local citizens' actions prevented these plans from being implemented and led to the innovative Napa River Flood Protection Project. In 1998, voters approved Measure A to fund this ecologically oriented, long-term approach to minimizing flood risk while maximizing the health of the river.

In the early 21st century, the Napa River finds itself in the midst of an unprecedented transition. For two centuries, the river has been increasingly

Napa River views. Juvenile Chinook salmon (top) collected from the Napa River at Yount Mill Road. The river at the Napa River Ecological Reserve near Yountville (bottom). *Top: photograph by Jonathan Koehler, April 1, 2007. Bottom: photograph November 28, 2009.*

The streams were stocked with fish of many kinds.

— VALLEJO, 1836[6]

modified and restrained. For decades, the river received the untreated sewage of growing cities; pollution from tanneries, wineries, and other industries; and assorted garbage.[7] Water-quality controls have improved and, in the past decade, large-scale restoration projects have been initiated in the river's lower reaches. Local farmers are improving a 4.5-mile reach in the Rutherford area, and restoration plans are being developed for the 9 miles between Oakville and Oak Knoll.[8] Barriers to fish migration are being removed, and the presence of steelhead and salmon has been well documented in recent years. The Napa River Trail, Trancas Crossing Park, and other projects are expanding community access to the river.

Nevertheless, the river still shows signs of distress. The streambed has cut down precipitously in many places, causing the failure of banks and adjacent property loss. Fine sediment often blankets the few potential high-quality spawning areas; the Napa River is still listed as impaired by sedimentation under the Clean Water Act. Increasing local water demand may threaten dry season flows. While the Napa River has not been transformed to the extent of many other California streams, it has been significantly altered, losing much of its complexity in the process. This chapter documents the trajectory of the last 200 years, peeling back the physical modifications to see the river before these changes.

The first three sections document major components of the historical river—wetlands, sloughs, and the floodplain; the next four explore trends in channel alignment, channel depth, sediment dynamics, and dry season flow. The final sections look at the ecology of the past and present river, how its character was controlled by valley topography, and its potential to recover former functions.

Rowing up the street. "This was the scene at Pearl Street in Napa today" read the caption for this AP wire photo of flooding in downtown Napa, February 1, 1963. *Courtesy of the California Historical Society, San Francisco Chronicle Collection, CHS 2001.568.tif.*

RIVER OR CREEK?

Most watercourses in the Bay Area were historically called *arroyos*, or creeks, while the term *río*, or river, was reserved for larger systems such as the Sacramento and San Joaquin. Both terms have been used for the Napa River, suggesting that it falls somewhere between ideas of creek and river.

Some of the earliest Mexican maps show the Río de Napa, distinguishing the river from Arroyo de Napa (present-day Napa Creek), a tributary to the river in the town of Napa. However, another map of the same era refers to the Napa River as Arroyo de las Trancas. Also in the 1840s, Yount, Dana, and Clyman all referred to Napa (or "Napper") Creek.[9] After American settlement, federal surveyors Tracy (1858a, 1859) and Dewoody (1867, near Calistoga, and 1879, near Oak Knoll) recorded the Napa River; a land case called the river Río de las Trancas (Squibb 1861); and Loring (1853) even referred to it as the Caymous River, where it ran through the Caymus Rancho. Yet Bartlett (1854), Gray (1853), Thompson (1857), Kerr (1858), and later historian Adams (1946) all consistently referred to what we call the Napa River as Napa Creek. In 1860, the *Daily Alta California* acknowledged the nomenclatural complexity, reporting that "Napa river is not larger than many of the 'creeks' of the State, but it must be called a river to distinguish it from the Napa creek, which goes dry in the late summer." By the late 1800s the term "river" appeared to have, for the most part, prevailed.

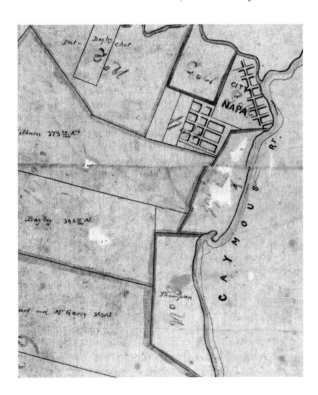 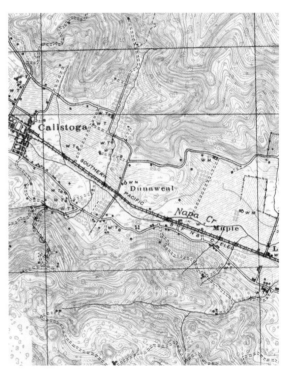

The Napa River was referred to as Caymous River in an 1853 map of Nicolas Higuera's tract (left). At right, the name Napa Cr[eek] was used south of Calistoga in a map based on U.S. Army Corps of Engineers surveys in 1915. *Left: Loring 1853, courtesy of The Bancroft Library, UC Berkeley. Right: USACE 1927, courtesy of the Earth Sciences and Map Library, UC Berkeley.*

In low and marshy tule grounds along the river, elk were found in abundance.

—MARY E. BUCKNALL, GRANDDAUGHTER OF LAND GRANTEE GEORGE YOUNT

THE RIVER SPREAD INTO WETLANDS

In the fall of 1852, a survey team followed the Napa River, using compass and chain to convert the meandering course of the river into an official cartographic and legal boundary. Just north of Dry Creek, the character of the river abruptly changed: the channel disappeared, leaving no dividing line to survey. Nathaniel L. Squibb, a professional surveyor since 1826 and the County Surveyor of Napa County, had encountered one of the Napa River's great marshes:

> *State what is the character of the channel of the Napa River at the place marked 'Tule' on Exhibit NLS no. 2.*
>
> The branch seems to sink and form a marsh and the Napa spreads.
>
> *Has the Napa River a distinct channel at that point or any channel at all?*
>
> I examined the place and could not find a channel at that point similar to the channel above or below.[10]

Further north, near Zinfandel Lane, the main channel similarly downsized as it spread into several smaller sloughs winding through extensive "swamp and tule."[11] In these reaches, the river channel became less prominent as it spread into tule marshes and willow swamps, some over 100 acres in size. The river's wetlands contributed an array of valuable functions, from storing surface water and fine sediment to providing habitat for diverse wildlife. They likely provided high-productivity rearing habitat for steelhead and salmon.[12]

With no clear channel to define his course, Squibb abstracted a series of straight-line segments through the wetlands, until he rejoined the well-defined river channel on the other side. These anomalous segments in the otherwise sinuous boundary remain visible today in USGS maps.

The conversion of these complex systems to a single channel often took place quickly following American settlement, and with little documentation. In Calistoga, however, a local historian provided one of the best descriptions of the transformation of a riverine wetland complex:

> Before the Napa Creek [River] channel was formed, realizing that there should be a main one, John McFarling and John Cyrus, together with some other men, plowed large furrows where the channel now is. The heavy rains caused the furrows to deepen with the ultimate result that we now have—quite a creek running through our little city.[13]

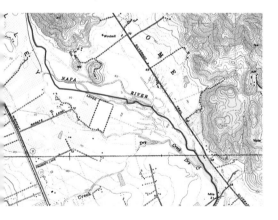

Squibb's river course. A straight line segment of the Yahome land-grant boundary established by Nathaniel Squibb is still shown (red line) on the contemporary USGS quadrangle, revealing the location of the former marsh.

Wetlands along the river. Basket sedge and other wetland species (above) occupy the floodplain between the Napa River and Conn Creek at the Napa River (Yountville) Ecological Reserve. An 1866 county survey (left) shows floodplain sloughs from the Napa River spreading into extensive "Marsh Land" **a** just north of Yount's Mill (shown at lower right). The survey, for the realignment of a road around the marsh, also identifies the Mill Race **b** that carried water from the river to the mill. *Photograph November 2009. Map Pierce 1866, courtesy of the Napa County Surveyor's Office.*

ISLANDS AND SLOUGHS

Two hundred years ago, the valley had islands, which were surrounded by water, and sloughs, far from the tides. Both were common features, and sometimes even landmarks.

At least 32 miles of floodplain sloughs (also called overflow, backwater, secondary, or side channels) branched from the main stem[14] of the Napa River, ran roughly parallel, and rejoined the main stem downstream. Ranging from 10 to 50 feet wide,[15] these features extended the amount and diversity of both aquatic and riparian habitat well beyond the primary river channel. They also provided additional channel capacity during high flows.

The sloughs, or side channels, branched through mosaics of wetlands, riparian forest, and seasonally flooded meadows, defining distinct "islands" (the largest, near Yountville, was over 300 acres). Early immigrants to the valley testified about these features, when questioned about their memories of the 1840s to determine conditions at the time of the land grants:

> "I know 4 or 5 islands in the Napa River."—Salvador Vallejo

> "I am well acquainted with the whole country and the island as laid down on the Exhibit but know of no name for the Island. I have always heard it called 'the Island.'"—Nathan Coombs

> "I know a place called Page's Island that I recollect of. I know of two places called the Island."—Elias Barnett [16]

These and other 19th-century sources confirm the branching character of the pre-modification Napa River. There were many channels running along the valley floor that were neither true tributaries nor the main stem river itself, engendering some confusion:

> *Is there any other stream except said 'dry creek' [Conn Creek] and Napa River which flow nearly parallel to each other, from a point in the valley nearly opposite Mr. Bale's house and passing down the valley below Mr. Yount's house?*
>
> I know of no separate stream, but there are sloughs running along and nearly parallel with said stream.
>
> *Do you mean that the sloughs commence nearly opposite Bale's house and run continuously down the valley, passing Mr. Yount's house.?*
>
> The sloughs I allude to are waters that primarily belong to the Napa River and have no separate heading to my knowledge.[17]

Many of the sloughs described by 19th-century residents and mapped by early surveyors were similar to the sloughs captured in early aerial photography, suggesting that the features were relatively stable. The early vintage of evidence, and consistency through the historical record, indicates that the sloughs were not the result of early land-use impacts (e.g., excessive erosion due to early grazing).

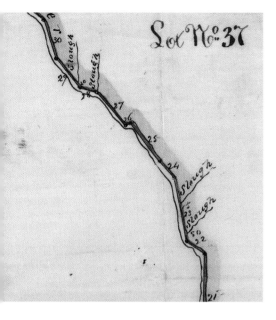

Napa River's sloughs. (top) A confirmation survey for the Rancho Yajome land grant shows four sloughs entering the Napa River south of Oak Knoll. *Tracy 1858d, courtesy of The Bancroft Library, UC Berkeley.*

Wood duck. (bottom) The elegant waterfowl were found along the Napa River's sloughs, where they could nest above stretches of slow-moving open water. Birder E. L. Bickford described seeing the formerly abundant species return to Napa Valley for the first time in decades during "the spring of 1925, when a flock of twenty-six was reported to be wintering in a small lake-like slough near the 'Little Trancas.'" After being "scattered by poachers," they returned to the "little wooded lake" three years later. Today wood ducks are found uncommonly in the valley.[18] *Photograph ©John White.*

Particularly in the lower half of the valley, Napa River's floodplain sloughs contributed to the productivity of the river. Given the surrounding permanent wetlands, some sloughs likely had perennial water. Slow-moving, well-wooded backwaters supported wood ducks and beavers, while providing rearing habitat and high-flow refuge for juvenile salmonids and other native fishes.[19] As these side channels have been filled in and disconnected, the main channel has had to carry more water, causing it to change as well.

An old slough. (left) Many disconnected sloughs are scattered throughout the valley. Some of these "lake-like" segments of former sloughs are currently used as agricultural ponds, long after the rest of the channel has been filled. This elongate, oak-lined pond might be the "little wooded lake" visited by Bickford when it was still connected to the river in the 1920s. The old grapevines in the foreground occupy the former slough course, filled over a half century ago. *Photograph December 4, 2001.*

Downtown slough. (below) This slough ⓐ paralleled the Napa River for 2 miles and rejoined the river at its confluence with Napa Creek. The slough's mouth can still be seen in downtown Napa and will be partially reestablished as the Oxbow bypass, an element of the Napa River Flood Protection Project. The cumulative length of the Napa River's historical side channels nearly doubled the length of the river.[20] *Sanborn Perris Map Company 1886, courtesy of the California State University Northridge Map Library.*

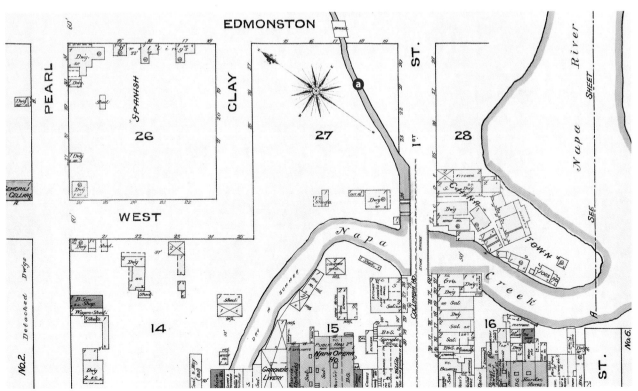

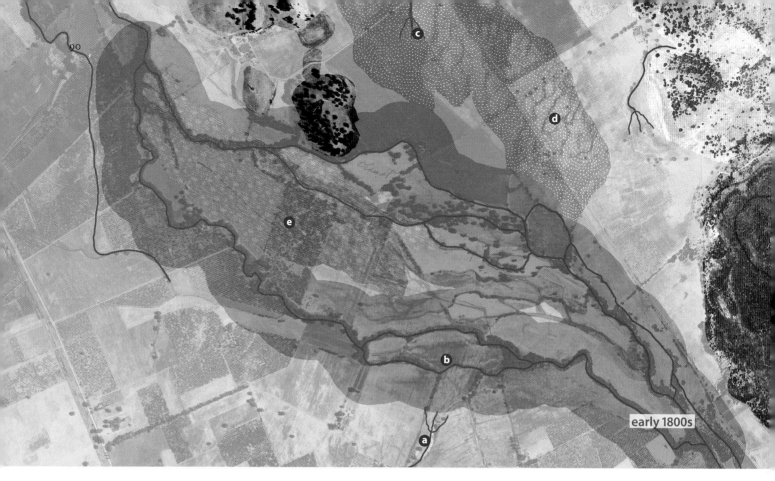

A disconnected slough in the Oakville area. *Photograph December 30, 2006.*

Floodplain sloughs south of Yountville. One of the most extensive channel complexes lay south of Yountville, where the floodplain broadened between the Yountville Hills and the Dry Creek fan. As shown in the early 1800s reconstruction (above), the river branched into a network of sloughs through freshwater marshes and willow swamps. The tributary Dry Creek **a** spread into "the Willows" **b**, while a small tributary from the north **c** dissipated its flows into vernal pools and swales **d**. The freshwater marsh encountered by surveyor Squibb is shown at **e**. By 1942 (opposite page, top), ditches were constructed, wetlands drained, and willow thickets cleared; yet the multiple channel system was still visible **f** and probably active in high flows. The largest side channels remain visible today (opposite page, bottom), but the branch to the south has been disconnected **g**; flows on the main channels are now constrained by levees. Some former channels are visible as ghost images in vineyards: faint darker green lines reflecting differential soil characteristics in the former sloughs **h**. While most of the river's sloughs have been disconnected, some of them could potentially be reconnected.

	Valley Freshwater Marsh
	Wet Meadow
	Vernal Pool Complex
	Broad Riparian Forest (with historical channels)
	Oak Savanna
	Stream
	Distributary

1,000 feet

1942

2009

Reported flood extent, December 1955. (opposite page) This map overlays the area of flooding in December 1955 on contemporary aerial imagery. The floodwaters followed the historical channel network and extended laterally across the floodplain, including most of the area of former freshwater marshes. The boundary of the flooded area also corresponds quite closely to the foot of alluvial fans, showing how the floodplain is shaped by the adjacent alluvial topography. In this event, about 11,000 acres, or 25% of the valley upstream of the tidal marshlands, were flooded.

Alluvial fans define the valley floodplain. (below) This view looks at the 1955 flood extent and the location of perennial freshwater marshes in relation to valley topography. The vantage point is the hills on the east side of the valley near Lake Hennessey, looking west toward the Bear Canyon Creek and Sulphur Creek fans. The floodplain narrows where fans protrude and widens between the fans where large mashes formed. These subtle topographic variations are almost imperceptible on the ground.

- Extent of 1955 flood
- Historical freshwater marsh
- Alluvial fan (approximate)
- Historical channel
- 5-foot contours

THE VALLEY IN FLOOD

Valley wetlands, sloughs, and seasonally inundated lowlands all became connected during the brief but critical times of flood. Native fishes are adapted to take advantage of these pulses of productivity in the valley's lowlands, when broad, shallow waters supported a profusion of edible invertebrates. While fisheries restoration has tended to focus on access to headwater streams, recent research has suggested the importance of floodplain and wetland rearing to salmonids and other native fish. The intensive feeding opportunities provided by these off-channel habitats can enhance salmon growth rates, increasing their chances of successful marine survival and return as spawning adults.[21]

The Napa River has been heavily modified to reduce flooding, so it is difficult to visualize its former, unrestricted extent. After the destructive December 1955 flood, however, a detailed map was produced showing the area of inundation.[22] The 1955 flood was a major event, causing widespread damage throughout Northern California, so not all of the area shown by the map would necessarily be covered during smaller, more frequent floods. However, the flood extent corresponds to the historical distribution of floodplain sloughs and wetlands that would be expected to flood on a regular basis. The map suggests the broad historical extent of off-channel and floodplain habitat in a major event.

Because so little floodplain function exists along the Napa River today, the restoration of relatively small areas (compared to the historical extent) could significantly increase ecosystem functions.[23] Where possible, strategic floodplain restoration should focus on existing flood-prone areas, which may also be less productive and less valued for agriculture and other land uses.

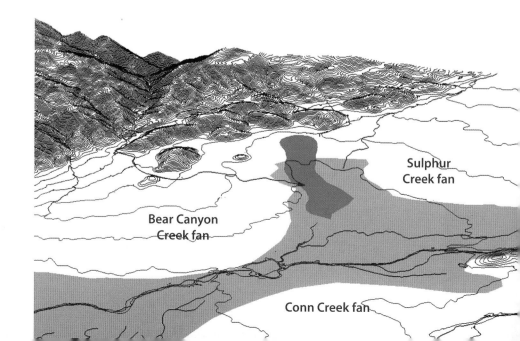

HAS THE RIVER BEEN STRAIGHTENED?

Compared to prominent California rivers such as the Sacramento and San Joaquin, the Napa River's main stem appears relatively straight. Given the prevalence of straightened, channelized rivers throughout California and the U.S.,[24] one might wonder whether this pattern is natural or artificial.

Certain reaches of the river have been straightened; former meanders can be identified from historical maps and aerial photography. However, early documents more commonly show close correspondence between historical and contemporary alignment.[25] The river has been straightened in places, and side channels removed, but the overall shape of the main stem has not dramatically changed.

While highly sinuous, meandering rivers are more well recognized and perhaps culturally preferred,[26] the Napa River was naturally fairly straight, with no oxbow cutoffs and a low sinuosity (1.1–1.2)[27] for most of its length. (The large meander in Napa known as "the Oxbow" was sufficiently unique to justify its name.) Only in its tidal portion was the river substantially more sinuous. Not coincidentally, this reach has experienced the most straightening (see the "Dredging the Tidal River" section on page 130).

With prominent secondary channels and islands, the Napa River exhibited a branching channel pattern, also known as anastomosing or island-braided.[28] Rather than having a highly sinuous form with oxbow cutoffs and lakes, the Napa River's complexity came from its wetlands, floodplain sloughs, gravel bars, fallen logs, and beaver ponds. While the Napa River reflects many impacts, straightening, for the most part, is not a major one.

Napa River sinuosity. The river was most sinuous in its tidal reach, which is where the most severe straightening has taken place (to facilitate navigation). While there has been some loss of meanders in other reaches, overall sinuosity has not decreased appreciably. Except for its tidal reach, the historical and contemporary river would be classified as "straight" or "low to moderate sinuosity."[29]

Reach	Historical Sinuosity	Modern Sinuosity	% Change
Butler Bridge to Trancas Street (tidal)	1.45	1.25	−14%
Trancas Street to Dry Creek	1.22	1.15	−6%
Dry Creek to Conn Creek	1.20	1.20	–
Conn Creek to Bale Slough	1.19	1.18	−1%
Bale Slough to Sulphur Creek	1.23	1.17	−5%
Sulphur Creek to Ritchey Creek	1.18	1.18	–
Ritchey Creek to Blossom Creek	1.14	1.13	−1%
Full Fluvial River (Trancas St. to Blossom Creek)	**1.21**	**1.19**	**−2%**

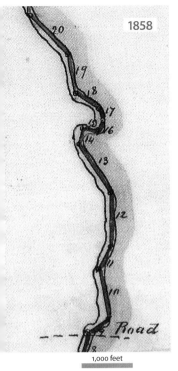  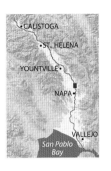

Channel plan form, 1858–2009. (left) Comparison of contemporary aerial photography to an 1858 survey shows that although one prominent meander no longer exists, the general shape of the river has remained similar. This example is in the Trancas Street to Dry Creek reach, which has experienced a 6% reduction in sinuosity. *Far left: Tracy 1858d, courtesy of The Bancroft Library, UC Berkeley.*

Channel plan form, 1942–2009. (below) In this reach near Rutherford, a sinuous and narrow river course has been converted into a much straighter and deeper channel.

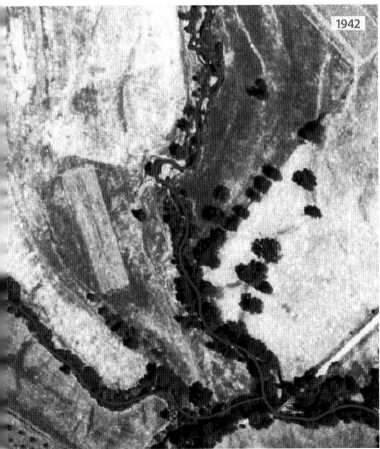 

HOW DEEP WAS THE RIVER?

In much of the West, rivers have experienced severe incision over the last century, cutting down deeply into their floodplain.[30] The change can be imperceptible from year to year, but represents a major shift in how the river functions and in its influence on the surrounding valley. Heavily incised streams drain groundwater, lose access to floodplain habitats, and erode adjacent property. On the Napa River, there has been anecdotal observation of general downcutting during the second half of the 20th century, but there are no long-term quantitative data sets.

Quantitative evidence for these kinds of vertical changes can be difficult to find, since multiple sources of relatively precise information are required. To assess changes in the river's bed elevation, we compared fourteen historical measurements of channel depth to contemporary river cross sections.[31]

The most striking evidence for the downcutting of the Napa River comes from Napa's iconic old bridges. Early-20th-century photographs and engineering drawings were compared to present-day conditions. These locations were precise and easily identified. A complementary source of information was provided by California Department of Fish and Game biologist C. K. Fisher, who described fish habitat conditions along the entire main stem, including channel depth, in 1959.[32]

In combination, these sources confirm a general trend of incision during the 20th century. They suggest that 6 to 10 feet of incision has been common over

Zinfandel Lane Bridge. The bridge was originally constructed with a hidden concrete footing below bed level that supported the stone pillar **a**, as can be seen in the historical photograph. Since construction, the footing has been exposed by channel erosion **b**, necessitating the pouring of a new, lighter colored concrete slab **c** to protect the bridge footers. Incision of 4 to 5 feet is estimated based on the height of the exposed footing. *Photograph (below left) courtesy of Al Edmister. Photograph (below right) by Sarah Pearce, October 2006.*

the last 40 to 100 years, which is in the range of previous estimates.[33] They also show substantial variability, presumably reflecting the temporal and spatial complexity of channel processes, as well as measurement error. Channels do not incise uniformly but reflect local variations in bed material (cutting more slowly through bedrock and hard clay surfaces), gravel supply, bed/bank stability and stabilization efforts, and the relative proximity of larger watershed processes.

In fact, because many of the historical data focus on bridges, which tend to be built in relatively stable areas and are often actively protected from erosion by concrete aprons or riprap, they may be biased toward showing less incision. Also, these data do not address potential earlier changes. Qualitative 19th-century information about channel depth, such as the use of the river for agricultural irrigation and to power gristmills, also suggests a significantly shallower channel than observed today.

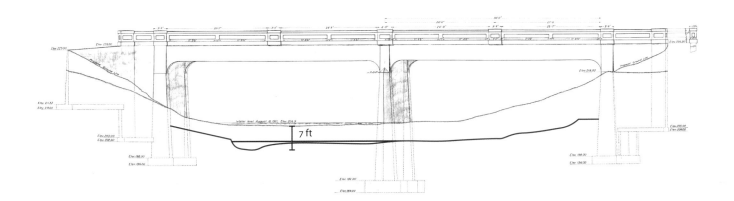

Pratt Avenue Bridge. A 1921 construction drawing[34] (above) helps document changes in channel dimensions. The architectural details in the drawing can be identified at the site (shown in the photograph at left), providing a frame of reference for the measurements. Based on these data, we can see that the channel has incised approximately 7 feet since the bridge was built. The right bank has also undergone a significant amount of erosion, although this is likely a localized change. *Construction drawing courtesy of the Napa County Surveyor's Office. Photograph by Sarah Pearce with Jonathan Koehler and Paul Blank of the Napa County Resource Conservation District, October 2006.*

DISAPPEARING GRAVEL BARS

Much of the contemporary Napa River consists of long stretches of deep water, called glides or run-pools. Researchers have speculated that the earlier river had a more complex channel morphology—sequences of large gravel bars, deep pools with shallow margins, and riffles (shallow flowing water)—that reflected the natural dynamics of sediment and water on the river.[35]

Historical evidence supports this concept. A number of early-20th-century landscape photographs show broad, unvegetated gravel or sand bars bordering pools of shallow or deep water, indicating the kinds of frequently rearranged and scoured gravel beds required for salmon spawning. Similar characteristics were widely distributed along the river, as seen in early aerial photography from a range of years. Comparison with contemporary imagery shows a dramatic loss of these features in recent decades.[36]

As the main channel has downcut, adjacent gravel bars that were formerly exposed to frequent disturbance by high flows have been left comparatively high and dry. This has allowed the extension of riparian vegetation into the channel, further restricting the movement of gravel by floods.[37] In addition, gravel mining and the construction of dams on Conn, Kimball, Rector and other creeks have reduced the natural supply of gravel to the river.[38] The greater erosive force associated with more confined flows also tends to prevent the formation of new bars. While current enhancement projects look to artificially introduce gravel to the river, 19th-century accounts marveled at the tremendous amount of gravel carried through the river into its tidal reaches:

> Lieutenant Potter said that he could not understand how such an amount of heavy gravel could be brought down. 'I guess you never saw one of our floods or you would not be surprised at all,' said Editor Francis.[39]

> The river is subject to considerable freshets, which bring from above floating trees and detritus, composed in part of gravel, which settles to some extent in the channel and forms shoals.[40]

As natural gravel transport has been restricted, increases in fine sediment from hillslope and bank erosion have resulted in an imbalance between fine and coarse sediment. Reestablishing a more balanced sediment regime that produces dynamic patterns of bar, riffle, and pool is a central challenge to reestablishing a healthy river.[41]

Summer activity on a Napa River gravel bar. A woman washes clothes with a washboard in the riffles while several people sit on the opposite shore. The relatively open, shallow river banks once provided people with easy access to the river. *Courtesy of Sharon Cisco Graham.*

Lower Napa River, 1942 and 2009. (above) In this reach, as in others, large gravel bars evident during multiple decades of the 20th century have now been colonized by riparian vegetation that encroached into the formerly more open channel, probably as a result of channel incision. These views along Silverado Trail south of Oak Knoll Avenue also illustrate the loss of the Napa River's side channels. The partially intact courses of side channels can be traced on both sides of the river in the 1942 image; elongate agricultural ponds occupying former channels can be seen in the contemporary image.

Companion views of the Napa River at Pope Street Bridge in St. Helena, June 5, 1907. The downstream view (opposite page, top) shows a complex channel morphology that would have supported a diverse range of native fish within a small area. Sacramento suckers, tule perch, and Sacramento pikeminnow would use the shaded deep areas along the undercut banks; juvenile salmonids, suckers, and pikeminnows would find protection in the shallow fringe habitats along the bar edges. The view from the opposite side of the bridge, looking upstream (opposite page, bottom), reveals a very different setting. A massive bar (probably due in part to the obstruction of the bridge) and a large root wad from a fallen tree attest to the powerful processes shaping the river. *Photographs by Turrill and Miller, courtesy of the Society for California Pioneers, catalog nos. C030445 and C030446.*

A swimming hole north of St. Helena, 1949. *Photograph courtesy of Tom Wilson.*

DID THE RIVER FLOW YEAR-ROUND?

Since most coastal California streams have no snowpack and experience an annual summer drought, the presence of water in the dry season is not guaranteed. Given the long dry season, many Bay Area streams naturally went dry in some reaches, even before diversions.[42] Presently the Napa River exhibits little or no flow in some reaches during late summer/early fall, especially in dry years.[43] But the nature of prior stream flows is often not obvious: some California streams have more summer flow than they used to (because of water imports and/or reservoir releases), while others have less (because of diversion to agriculture and/or cities).[44]

The available historical evidence, however, does indicate the reliable presence of summer water in the Napa River. While some local creeks were described as dry or *seco* in historical accounts, no known sources described the Napa River this way. On the other hand, dry season flows were modest: even before diversions, the Napa River receded to a trickle each year. Base flows naturally increased downstream due to percolation from valley groundwater; the river was a "gaining" stream—that is, it intercepted groundwater rather than contributing to it.[45]

For the lower half of the non-tidal portion of the river—the 15 miles between St. Helena and Trancas Street—historical records suggest there was perennial surface flow. Valley resident Bartlett Vines, who testified in the Caymus Rancho land case in 1861, was asked to compare the "dry bed" of Conn Creek "in the dry season" with "the bed of Napa River at the same season." Vines responded that, in contrast to Conn Creek, "Napa River is considerably larger with deep banks and water running in it." Surveyor Thompson stated in 1857 that the Napa River "affords in many cases opportunities for irrigation" within the Caymus Rancho (approximately Zinfandel Lane to Yountville Cross Road)[46]—an application that would require available water in summer and fall. Just upstream of Trancas Street, the Napa City Water Company built a diversion dam on the Napa River that served as a dry season water source for the city from the 1880s to 1906.[47] Early hydrologists measured flow at 2.7 cubic feet per second at Oak Knoll on August 28, 1911, before significant diversions.[48] These descriptions and practical uses each suggest a modest but reliable dry season flow. In addition, the presence of extensive freshwater marshes and willow swamps along the river between Rutherford and Oak Knoll indicates water was at or near the ground surface year-round, which would be consistent with perennial base flow.

Upstream of St. Helena, flow was more restricted and spatially intermittent. A statewide survey of flow characteristics in 1910–11 included several measurements on the Napa River and tributaries.[49] These data show the steep seasonal recession in flow, resulting in little flow in Calistoga, at Bale Lane, and at Pope Street in August 1910. Nevertheless, some flow was present. Similarly, in 1936 Grinnell reported that water in the upper river was "very low, scarcely moving" in an average late November with no preceding rain.[50] Despite the low flows, though, the most famous swimming holes on the Napa River were found in these upstream reaches, probably because of perennial tributaries such as Mill and Ritchey creeks. Wilson's Swimming Hole was just south of the Bale Lane Bridge; a series of swimming holes were named for locals McCaffery, Laurent, and the Moraff Brothers.[51] The upper river may have had little surface flow at certain times and places, but it was not completely dry. It maintained deep perennial pools through the dry season that could support steelhead and other native fish.

While the river and its native fauna are naturally well adjusted to a Mediterranean climate, these systems are inherently sensitive to reductions in water supply. It appears that the river may be more prone to going dry earlier in the year now than it was historically.[52] This is not unlikely given the impacts to the watershed. Conn Creek, now dammed for urban water supply, once contributed substantial subsurface flow to valley groundwater and the river.[53] Near-surface groundwater, now rapidly drained in the winter, used to percolate slowly to the river. Dry farming, practiced into the 1970s, has given way to irrigated agriculture. With climate change and potentially increasing agricultural water needs, strategic design and management of the water budget will be critical for the persistence of river functions.[54]

Stream flow measurements, 1910–11. These data, from a USGS assessment of "Water Resources of California," show the progressive increase in flow downstream on the Napa River, as the river receives input from groundwater and tributaries. The data also show the decrease in flow at all sites through the course of the spring and summer. The measurements were taken before significant dams and diversions, when the valley was primarily dry-farmed. (The water year 1909–10 was about 15% below the long-term average for water flow, while 1910–11 was 5% above average.[55] Units are cubic feet per second (cfs). Parenthetical phrases are the authors' site descriptions.)

Location	May 24–27, 1910	July 12–13, 1910	August 13–14, 1910
Lincoln Avenue ("bridge at Calistoga")	2	0.3	0.1
Bale Lane ("100 feet below bridge on road from Bale railroad station, 4 miles north of St. Helena")	5.1	1.5	0.4
Pope Street ("75 feet below bridge, 1 mile northeast of [St.] Helena")	5.5	1.4	0.4
Rutherford Road ("bridge, 1 mile east of Rutherford")	11	2.7	1.4
Oak Knoll Avenue ("600 feet below second bridge, 6 miles northwest of Napa")	—	—	2.7*

*The single Oak Knoll measurement was taken during the following year, on August 28, 1911.

RIPARIAN FOREST WIDTH

Today much of the Napa River is bordered by thin strands of riparian forest, 50 to 100 feet in width. While some riparian functions can be provided by such narrow corridors, most functions require substantially wider zones. Research suggests that much wider ones—100 to 500 feet or more on each side of the channel—are required for many riparian wildlife species. Broad riparian forests can provide nesting sites for the wood duck and yellow-billed cuckoo, cover for elk and deer, and food and building materials for beavers.[56] How wide was the Napa River's riparian corridor?

Direct early evidence is limited, since riparian forest was rarely explicitly mapped. But several lines of evidence indicate that the river historically supported a much wider riparian forest than exists today, at least on some reaches. On a 4-mile reach of the river upstream of St. Helena, 1942 aerial imagery shows a series of five apparent riparian forest remnants 200 to 400 feet wide. These patches, each more than 800 feet long, stopped and started at property or field boundaries, suggesting that they represented residual features of a formerly continuous corridor. An 1873 map of the Mill Tract (east of Calistoga), one of the few to explicitly show riparian forest, affirms a similar forest width continuing for a mile between the remnants visible in aerial photography, supporting the interpretation of a more broad and continuous forest.[57]

On the lower Napa River near Zinfandel Lane, GLO surveys by Gray in 1853, Thompson in 1857, and Dewoody in 1866 all documented even wider,

early 1800s

1942

2009

Riparian forest loss on upper Napa River. A number of broad riparian forest areas are visible along the Napa River in 1942 imagery. In many places, they terminated abruptly at the boundaries of fields, suggesting prior clearing. Using the 1942 evidence and other data, we can estimate the earlier historical extent compared to the reduced contemporary extent.

500 feet

willow-dominated riparian forests associated with side channels of the river. Dewoody recorded "willow thickets" several hundred feet wide (726 feet wide at a slightly oblique angle) on the northeast side of the river. On the other side of the river, he traipsed through a 528-foot-wide willow thicket perpendicular to the river and, after a 792-foot gap corresponding with a fence line (which had likely been cleared), he continued through another thousand feet of willows.

These are the best documented riparian forests along the river, but it is likely that there were more, as willow lands were especially valuable for agriculture and often subject to early clearing.[58] Coombs' property near Trubody Lane, which encompassed side channels of the river, was called "the Willows"; Tracy surveyed a "willow thicket" 1,400 feet wide in the vicinity. A tract of land north of Napa was known as *Sausal* (willow grove) Rancho.[59] An 1871 article suggested that willow lands covered substantial areas along the river and foreshadowed their clearing: "I cannot help thinking how much better it would be for all parties if land-holders would sell these waste lands [along the Napa River]…to such of those as would make a thorough business of clearing out the useless willows and covering the broad and fertile acres with blackberries, gooseberries, currents, or other fruits."[60] It is likely that broad willow forests like those sampled by historical data were common along the floodplain slough complexes of the lower Napa River.

In a valley with little other dense forest, the river's broad riparian forests provided unique ecological functions. While most of these features are long gone, they have shown the potential to recover, expanding rapidly when provided appropriate space and conditions. Restoration should look for appropriate places to bring back this missing element of the river.

Riparian forest and wetlands near Zinfandel Lane. Willow forest extended well beyond the Napa River main stem (here referred to as "Napa Creek"), as documented by mid-19th-century surveys. The independent transects by Gray (1853) and Dewoody (1866) showed a complex system of broad riparian forests ("swamp," "willow thicket"), tule marshes, and multiple channels ("slough"). The overlay on 2009 imagery shows the overall reduction in system complexity, although portions of at least one slough noted by Dewoody can be identified. The arrows on the map indicate the direction taken by the surveyor. The annotations are excerpts from the surveyors' field notes.[61]

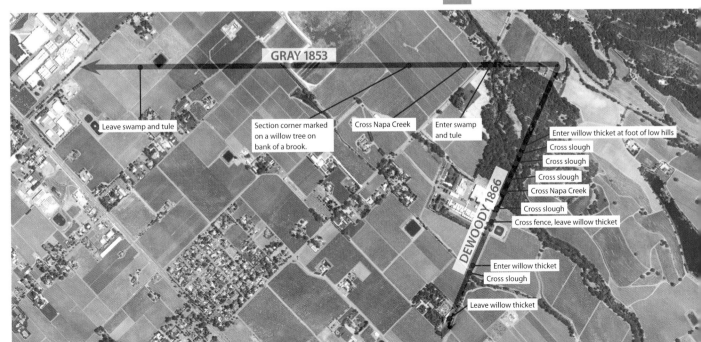

NARROWING OF THE RIPARIAN CORRIDOR

No direct evidence is available to comprehensively map the pre-modification extent of riparian forest along much of the Napa River. However, as seen in the previous section, a range of historical sources support the presence of broad riparian forests along the river. To estimate the historical extent, we measured the forest width as evidenced by these sources and extended these data to likely similar, but less well-documented, reaches. To assess long-term changes, we then compared this estimate to a recent map of contemporary riparian area produced by the Bay Area Aquatic Resources Inventory (BAARI).[62]

Apparent riparian forest remnants shown in 1940s aerial photography averaged 260 feet along several reaches of the river. GLO surveys indicated much broader riparian forest in the multi-channeled reach near Zinfandel Lane. We found the average distance from channel (main stem or slough) to willow thicket margin in these areas to be approximately 660 feet. We extrapolated these data to other similar reaches based on qualitative descriptions, air photo evidence, the presence/absence of multiple channels, and valley width. For example, we applied the evidence for 660-foot-wide riparian forest in the Zinfandel area to the other broad, floodplain slough/wetland areas near Yountville. This extrapolation was supported by textual descriptions of the Willows and of the general prevalence of willow lands along the river, as well as by Tracy's 1859 survey. Based on the aerial photo measurements of remnants (described above), a 260-foot buffer was applied to these reaches.[63]

The resulting map suggests the extent of riparian forest and illustrates how it varied along the river in relation to large-scale physical controls. The broadest forests, intermixed with floodplain wetlands, occurred where the valley is widest. Narrower corridors were found where the floodplain is naturally constricted by adjacent topography, such as Sulphur Creek's alluvial fan near St. Helena. This analysis suggests that most of the river channels were bordered by very wide forests, greater than 250 feet on each side of the channel.

Present-day mapping shows a dramatic narrowing of the riparian corridor, as broad forest reaches have become much thinner and many forests along side channels have been lost altogether. The removal of riparian forest translates into a loss of habitat for native wildlife, a deficit in the natural recruitment of wood to the channel, and bank instability.

Birds of the riparian forest. The first confirmed yellow-billed cuckoos (top) in California were collected in the Napa Valley in 1862; nests were found as late as 1902. The cuckoo's range has contracted throughout the West with the loss of broad riparian forests; it is listed as a California Endangered Species. The yellow-breasted chat (bottom) also used to nest commonly in forests along the Napa River and is now a Species of Special Concern.[64] *Paintings by Andrew Jackson Grayson, courtesy of The Bancroft Library, UC Berkeley.*

Change in the extent of riparian forest of different width classes along the Napa River. (right) Historically, the river was bordered by broad riparian forests, especially many miles of densely wooded side channels in the mid-valley. Today the overall river length has decreased, mostly because of the loss of side channels, and the remaining riparian corridors are comparatively narrow. Widths refer to the extent of forest on one side of the river channel; unless constrained by an adjacent hill, the full corridor width would be twice these values. Contemporary data from BAARI.

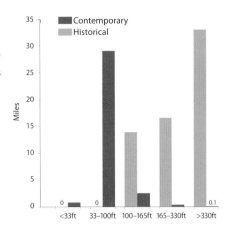

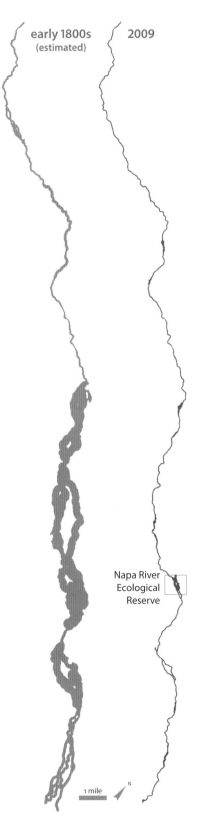

Riparian forest remnant. (above) The Napa River Ecological Reserve near Yountville maintains the river's largest remaining forest.

Riparian forest along the Napa River, early 1800s and 2009. (right) These maps compare riparian forest along the Napa River from Calistoga to Napa. There are a few wider contemporary areas (including the Yountville Ecological Reserve, shown above), but these are relatively small compared to the broad forests indicated by historical data.

BEAVERS AND SNAGS

While the presence of the river makes riparian forest possible, the trees in turn contribute to the structure and function of the river. In particular, fallen trees create hydraulic complexity both through the work of beavers and as large woody pieces that enter the river from erosion and old age.

Previous researchers have suggested that beaver were historically absent from the Napa River. Grinnell excluded the Bay Area in his estimate of former distribution,[65] and the account of John Work, one of the earliest trappers to come through the North Bay, has been read similarly. In the Napa Valley on April 9, 1833, Work wrote, "the little river where we are encamped at appears very well adapted for beaver yet there appears to be none in it." However, six weeks later Work suggested that beaver were in fact present in the river. The expedition apparently sent a side party to the Napa River, where they had earlier "found a few beaver." Work also reported beaver a few miles away in the Sonoma Creek watershed, making it highly probable that they were in the neighboring Napa River. Subsequent references also support the historical presence of beavers on the river.[66]

Trapping likely removed most beaver from the watershed by the 1840s. But before that time, beaver dams would have increased the extent and persistence of wetlands and ponds along the river even beyond the amount documented in the 1850s and 1860s, providing a range of ecosystem functions. By building dams from riparian trees, beavers create areas of temporary flooding that capture fine sediment, prevent channel incision, and enhance summer base flow.[67] The importance of beaver ponds to salmon populations and overall stream function has been increasingly recognized in recent years; their return can help speed river recovery.[68]

Beavers have recently been found in the Napa River for the first time in years—in the vicinity of Rutherford and downstream of Oak Knoll.[69] While they can be considered a nuisance because of their potential to forage in croplands and increase flooding, beavers have also been appreciated at times for raising groundwater and maintaining base flows.[70] In reaches with sufficient riparian vegetation and where their activities do not present unacceptable management risks, the return of these hard-working natural stream engineers could contribute to the restoration of the river.

Riparian trees are also essential as fallen logs and root wads, which wedge across the channel to help form pools, riffles, and gravel bars. Recent analyses have suggested that the river lacks in-channel wood, although there is little direct evidence of earlier conditions.[71] On larger rivers, records of snag removal (to maintain navigation) are often used to document the historical role of fallen wood in river functions.[72] These sorts of data are not

Gnaw marks. (top) Beavers are mostly nocturnal and rarely seen, but they leave substantial evidence of their activity. These cut trees were observed along the river near Rutherford. *Photograph by Jonathan Koehler, December 7, 2006.*

Beaver dam on lower Napa River.
(bottom) *Photograph by Jonathan Koehler, October 9, 2008.*

available for the Napa River except in its tidal reach, where Coast Survey and Corps of Engineers documents do mention the removal of snags—indicating the downstream transport of in-channel wood (see the Tidal Marshlands chapter). The *Napa Daily Register* even urged the state to station a "snag boat" on the river to remove logs and stumps.[73] Anecdotal evidence suggests that the historical prevalence of large logs and debris jams also extended further upstream.[74] The branching channel pattern that occurred in many reaches of the river was likely due to debris jams, which caused the channel to abruptly switch course.[75]

While the Napa River has not been converted to an engineered channel like many other lowland rivers in California, it has experienced some of the same effects. The channelization of wetlands, the expansion of the drainage network, the disconnection from side channels, the decline of beavers, the removal of riparian forest, and the loss of woody debris all have contributed to the rapid, unmodulated transport of water and sediment through the system. Common results of such modification are increased bank and bed erosion, resulting in a deeper and wider channel with less habitat diversity. Compared to the slower Napa River of the recent past—in which water spread and converged, sped up and slowed down—the present river is a straight chute, a water highway.

Snags are brought down by the winter freshets, and these also are hindrances to the navigation.

— U.S COAST SURVEY SUB-ASSISTANT AUGUSTUS F. RODGERS, 1858

American beaver. *Painting from* Fur-bearing mammals of California *(Grinnell et al. 1937), © 1965 by the Regents of the University of California. Reprinted by permission of UC Press.*

JEPSON'S NAPA RIVER

While maps and aerial photographs help reconstruct historical river alignment and riparian forest extent, we must turn to other sources to illustrate the riparian forest composition. California's preeminent botanist fills out the picture nearly single-handedly, through field observations scattered through four decades of field books, testament to Willis Jepson's lifelong love of the Napa Valley.

Jepson collected plants along the river at the Pope Street Bridge in St. Helena as early as 1892 and as late as 1936—on the way to botanizing Howell Mountain to the east. (The road also passed St. Helena Sanitarium, where, after a heart attack in 1945, he would spend some of the last months of his life.) The cumulative observations near the bridge create a rich picture: wetland species like water smartweed, nutsedge, and basket sedge (which formed "stools 1 to 2 feet broad and 2 to 3 feet high in the flood bed"); more xeric plants like California brickell bush on "Flats"; five species of willow (red, arroyo, shining, Scouler's, and narrowleaf); wild grape and a "fine cluster" of California (big-leaf) maple 45 feet high.[76]

Further downstream, in 1934, he stepped back for a broader view:

> Driving south in Napa Valley, one finds, a quarter of a mile north of Oakville… a wide open field stretching away to a screening border of splendid Valley Oak and tall Red Willow and Yellow Willow along the Napa River.[77]

Jepson also described Oregon ash and red alder (both more common further north), as well as thickets of California rose as much as a quarter mile long.[78] Intriguingly, he noted the absence of two common California riparian trees, Fremont cottonwood and California sycamore. In 1900, he commented in a field book, "so far as I have observed the cottonwood is not found in Napa Valley" and he expanded the point a decade later: "I have never seen the Fremont cottonwood nor western sycamore in the Valley as native trees."[79]

Earlier accounts generally reinforce and expand Jepson's picture of the river. Bartlett in 1852 noted willows, suggesting large, mature trees in a consistently moist environment: "In the midst of the valley winds a small stream, called Napa Creek [River], its course marked by the graceful willows that grow along its margin." GLO surveyor Thompson's 1857 description had a similar emphasis: "willows on the banks of the sloughs and Napa creek [River]." In 1849 Revere noted how the river's "banks [were] distinctly defined by a long line of willows and other trees of larger growth."[80]

"Napa Creek, near St. Helena, Calif."
Turrill and Miller took this photograph near Jepson's favorite collecting spot on the river. The view shows a gravel bar on the left and dense willows overhanging deeper water on the right. *Courtesy of Sharon Cisco Graham.*

Land surveys mentioned a sampling of other tree species along the river, including "alder" (near Trancas Street in 1856), "ash" (near Larkmead Lane in 1886), "oak" (near Imola Avenue in Napa in 1875 and Yountville in 1855), and "live oak" (at Yountville Cross Road in 1858). There is an 1897 record of redwood along the river "near Napa city." In 1932 Spring and Lewis described live oak/valley oak/willow communities on the lower Napa River. We found few pieces of evidence that contradict Jepson's absent trees: the identification of a "Cottonwood Tree" across from present-day Kennedy Park, and another description by surveyor Thompson: "on the bank of the creek [Napa River] a growth of cottonwood elm and oak timber." Avery reported that oak "grow thickest where they belt the course of the creek, and are there mixed with sycamore, alders, willows, and a plentiful undergrowth of wild vines and bushes."[81]

How different is today's river? The Napa River still supports a diverse, mixed riparian forest community. Most of the species reported by Jepson and others can still be found along the river, although some, such as California rose, are less common.[82] Cottonwood appears to have become much more frequent. While there is some evidence that the species was not completely excluded as Jepson asserted, it is unlikely that, given his extensive time in the valley, the tree was as widespread a century ago.[83] Sycamore is still not found on the river, an illustration of the climatic differences between the Napa River and streams draining into the opposite end of San Francisco Bay just 50 miles to the south, where sycamores were historically dominant.[84]

The expansion of cottonwoods likely reflects basic changes in stream hydrology, such as channel incision. Cottonwoods often occupy former gravel bars that now function as terraces, less subject to flood scour.[85] Other shifts may have occurred as well, but are not conclusively documented. For example, valley oak, the predominant Napa River tree today, was not particularly emphasized as a riparian component in early accounts; local author C. A. Menefee even left the tree out of his list of Napa County riparian trees.[86] Conversely, willows appear to have been more prominent in places, particularly along the multi-channeled lower river. The graceful willows of the 1850s riverbanks now may be more confined to the incised stream bed, where they are close enough to summer flow to germinate; oaks may have expanded on the now-drier banks. It is possible that a more xeric, incised river has led to general shifts in riparian composition.

Wild grapes. Riparian forests were the natural vineyards of the Napa Valley. Wild grapes climbing riparian trellises fed birds, bears, and people. *Photograph by Jake Ruygt.*

GLIMPSES OF THE TIDAL RIVER CORRIDOR

At that time [1854] the banks…were covered with a dense growth of alders and willows. Now [1881] wharves, tanneries and mills have taken their place, and the cleared banks of the river give it the appearance of a canal.

— L. VERNON BRIGGS, 1931

In the tidal reaches, riparian forest followed the river's natural levees downstream until submergence and exposure to salts prevented the growth of trees. This ecological transition can be seen in images of bridges and steamships on the picturesque tidal river. As a result of levee-building and channel modifications, the Napa River today is almost completely denuded downstream of 3rd Street in the city of Napa, but it supported substantial and diverse vegetation in the recent past.

A riparian tree downstream of 3rd Street. (right) While mature trees such as this willow at Kennedy Park are rare in this reach today, many young native trees have recently been planted on the eastern river levee as part of the Napa River Flood Protection Project. *Photograph November 2009.*

Pictures of the tidal river. (opposite page) The navigable reaches of the Napa River, particularly from Suscol through downtown Napa, are particularly well documented in postcards, paintings, and photographs. *Top to bottom: "Napa River and 1st Street Bridge," Darms 1908; "Scene on Napa River, Cal.," circa 1910, courtesy of Todd Schulman; "Napa, Cal., River Scene. Near Asylum," circa 1910, courtesy of Eric Erickson; "Mt. Tamalpais from Napa Slough," William Lewis Marple 1869, courtesy of the California Historical Society.*

Looking upstream along the Oxbow. This view toward the 1st Street Bridge shows substantial riparian vegetation because river-oriented industrial operations—such as the tannery, warehouses, and lumber mills— were located slightly downstream.

Landings in south Napa. Riparian trees also persist in this view (looking downstream). The steamer landings were located between Division and Oak streets.

Near present-day Imola Avenue. Further downstream, low vegetation bordered the channel opposite the Napa State Hospital. Occasional trees were indicated by Kerr (1858) and described in accounts: "overhanging trees" were a hindrance to navigation. [87]

South of Butler Bridge. In the broad tidal reaches, a high water table and occasional saline influence precluded trees, except on small islands or river deposits. (See pages 120–121 for a larger image.)

FISH ASSEMBLAGES

The river contains Steel-head, Rainbow and Eastern Brook trout, Striped bass and many other desirable fish.

—SOUTHERN PACIFIC COMPANY, 1896

Changes in the landscape are expressed in the contemporary assemblages of fish found on the Napa River. Sacramento perch and thicktail chub occupied the floodplain wetlands formerly found along the river. These native fish have been extirpated from the watershed. (Thicktail chub is extinct, due to similar elimination of wetlands throughout its range.) Steelhead and Chinook salmon are threatened by extinction. Tidewater goby, which likely lived in backwater ponds of the tidal marshlands, is not currently found in the watershed.

On the other hand, the vast majority of fish species native to the watershed remain present today, even if in small populations. Historical research by Robert Leidy has documented 33 fish species native to the watershed. Of these, at least 29 can still be found.[88]

Given the decades of impacts, it might be considered remarkable that descendents of species that have lived in this valley for thousands of years still persist among us. Surviving nearly intact despite numerous challenges, these native fish assemblages represent the potential for a healthy river ecosystem in the future.

Steelhead. (right) A female adult in Heath Canyon Creek, a tributary to the Napa River. Recent monitoring efforts suggest the population is stable and may be increasing.[89] *Photograph by Jonathan Koehler, May 2008.*

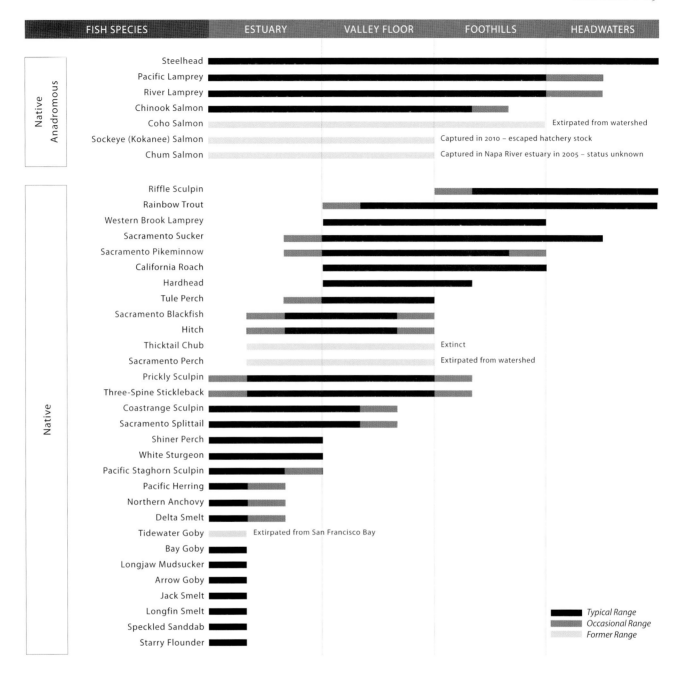

Native fish of the Napa River. Different fish occupied different parts of the Napa River and its tributaries, creating varying assemblages of species. This diagram, by Jonathan Koehler of the Napa County Resource Conservation District, uses black and dark gray bars to show which species are currently observed. Species no longer found are indicated in light gray. In addition to the native fish, 24 non-native species are now found in the watershed. There is no conclusive evidence for the historical presence of Chinook salmon, but they would be expected given the habitat characteristics of the Napa River and their presence in similar streams. It is unclear at this time whether the adult Chinook observed in recent years are of local or hatchery origin. There is also limited evidence for the presence of Coho salmon into the 1960s. Juvenile Chum salmon have been found recently and possibly were present in the historical watershed as well.[90] Note: Steelhead and rainbow trout are the same species; steelhead have a migratory life history while rainbow trout remain resident.

PATTERNS OF RIVER AND VALLEY

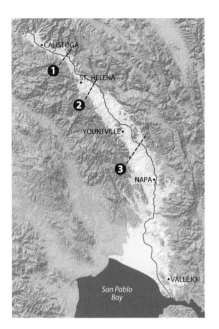

The different expressions of the Napa River derive largely from the basic topography of the valley—the physical shapes that control where water drains and collects. These subtle but influential patterns are perhaps the most fundamental characteristic of the valley; while we modify their surface, we rarely change their underlying form. As a result, contemporary cross sections of the valley correlate surprisingly well with its historical ecology, explaining characteristics of the river and adjacent habitats.

This spread presents several oblique views showing the surface topography of the valley, centered on the Napa River and looking upstream. Each view covers 1¼ miles in width and is based on recent Light Detection and Ranging (LIDAR) elevational data (actual cross sections derived from the raw data are shown below).[91]

Views of the river. The Napa River was a complex system, expressing different characteristics in different portions of its 35-mile course. The river corridor was more confined in the narrower upper valley and where broad fans (e.g., Sulphur and Napa creeks) provided lateral constraints on river movement. Between Zinfandel Lane and Oak Knoll Avenue, the river corridor was widest, leading to extensive side channels, wetlands, and floodplain complexes.

❶ **At Larkmead Lane,** riparian forest ⓐ occupied a zone several hundred feet wide adjacent to the channel, bracketed by gently sloping alluvial fans (of Ritchey and Dutch Henry creeks) that supported oak savanna ⓑ. The natural drainage of the fans prevented large wetlands from forming.

❷ **Near Zinfandel Lane,** the river was naturally confined by a steep hill to the east but spread into tule marsh and riparian forests in the low area on the west, between the river's natural levee ⓒ and the adjacent alluvial fan ⓓ.

❸ **South of Yountville,** the Napa River exhibited a subtle but distinct natural levee on both sides of the channel—a zone (about 1,000 feet wide) of higher land along the river. Broad marshes and willow forests were found in the low-lying area on the west. Alluvial fans were distant from the river channel here, leaving a wide wetland area traversed by floodplain sloughs.

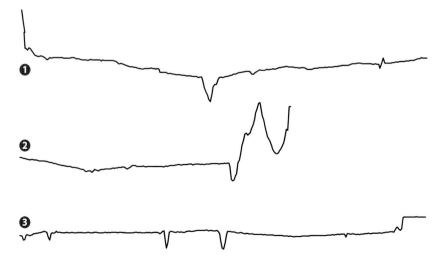

NAPA RIVER • 117

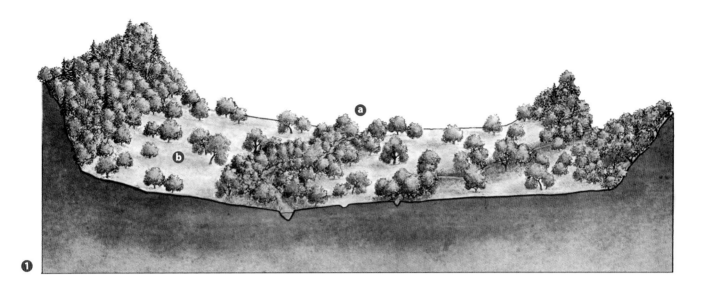

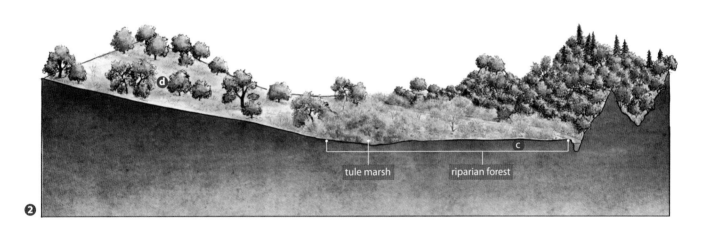

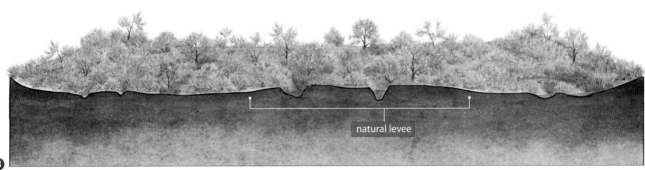

Illustrations by Jennifer Natali

THE RIVER RESPONDS

Rivers are dynamic systems, constantly responding to surrounding conditions. After a significant phase of incision, as the Napa River has experienced, riverbanks often become unstable and begin to fail, gradually producing a wider channel with an inset floodplain. Sequential aerial photos show this process in places on the Napa River. For example, at one site the main channel has eroded laterally as much as 300 feet in less than 70 years, carving out a broader channel and establishing new riverine features. At other sites, riparian forest has returned where 20th-century agriculture had previously encroached upon the channel. But for the most part, the river remains confined by surrounding land use.

Strategic rededication of land along the river to riverine processes can reduce the unpredictable erosion of banks and restore some of the missing elements of the river. Landowners in the Rutherford and Oakville reaches are beginning to do this. Reestablishing a healthy river will require a complex combination of actions, including giving land back to the river in places, restoring the balance between coarse and fine sediment, linking the river to healthy riparian forests and other floodplain habitats, and adapting the water drainage and storage network to be more supportive of riverine functions.

The river presents a tremendous opportunity: a tree-lined stream with little straightening, dominated by native trees, still supporting steelhead runs and nearly intact native fish assemblages. Where we have offered a modest margin between the river and intensive land use, we have often seen substantial recovery. The river will continue to be a dynamic force in the valley, responsive to our actions.

River recovery. In this reach just downstream of Oakville Cross Road, there is currently more riparian forest, gravel bar, and channel complexity than in 1942. Bank erosion has created a wider channel with an inset floodplain. Such accommodation is possible when banks are not hardened with riprap, and are allowed to retreat as erosion occurs.

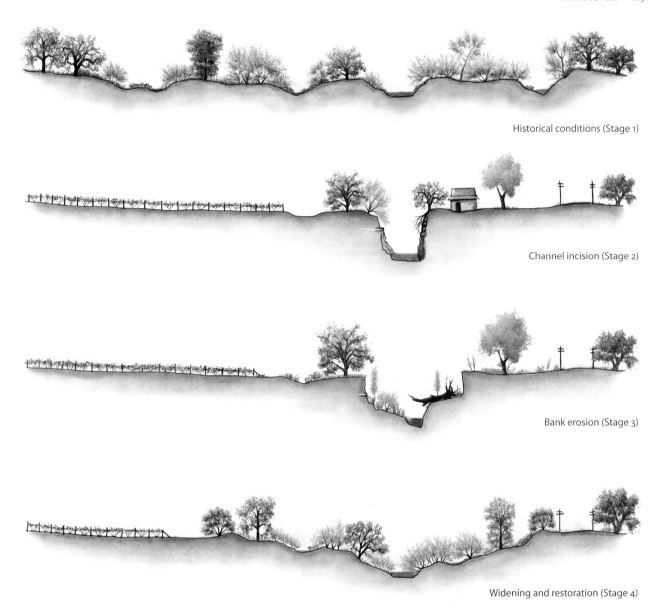

Historical conditions (Stage 1)

Channel incision (Stage 2)

Bank erosion (Stage 3)

Widening and restoration (Stage 4)

River evolution. These conceptual illustrations of the Napa River through time adapt the Schumm channel evolution model based on local historical evidence and current observations.[92] Stage 1 depicts the state of several reaches of the Napa River before 19th- and 20th-century modification, when the river was a shallower and broader channel, often with side channels and associated floodplain habitats. In the 20th century, the leveed, managed channel responded to increasingly concentrated flows by eroding a deeper, more uniform channel, shown conceptually in Stage 2. Today, the channel is widening and aggrading in a few places, often where banks are allowed to retreat, as depicted in Stage 3. This process creates a new inset floodplain with greater habitat complexity and could be enhanced depending on how the river is managed. Stage 4 imagines a widened and complex reach with regeneration of riparian vegetation and old side channels reengaged, which is the goal of several current restoration projects. The Napa River will continue to evolve and change depending on climate, land use, and management. *Illustrations by Jennifer Natali.*

Napa Valley's tidal marshlands before reclamation. An 1869 painting by William Lewis Marple provides a rare view of the marshlands. The view looks south toward Mt. Tamalpais as a steamer follows the sinuous route to the port of Napa. While the peak's height is exaggerated for dramatic effect, the image illustrates marsh features also documented by other sources, such as tall tules (towering over ducks in the foreground) and small islands with willows. *"Mt. Tamalpais from Napa Slough," courtesy of the California Historical Society.*

6 • TIDAL MARSHLANDS

Further south was the bay; and white sails of little schooners, outlined with a glass, appeared to split the salt meadows open, as they crept toward the little town of Napa.

—W. C. BARTLETT IN THE OVERLAND MONTHLY (BELDEN 1872)

Today most people approach the Napa Valley from the south by the highway. But 150 years ago, the best way was by boat, passing first through the broad tidal marshlands along the Napa River. From San Francisco, steamships entered the mouth of the river at Mare Island and headed in a generally northerly direction, skirting Slaughterhouse Point, Good Luck Point, Green Island, Bull Island, Horseshoe Bend, Lone Tree Reef, Carr's Bend, and Jack's Bend before reaching the landings at Suscol or downtown Napa. This serpentine, 15-mile course followed the natural deepwater channel created by the twice-daily inflow and outflow of the tides, continuously filling and draining hundreds of miles of tidal channels. Here the river was a broad tidal slough, closely connected to thousands of acres of surrounding marsh.

The common route up Highway 29 actually parallels the general tidal river course, just a mile or two away. But the freeway lies a world apart. Despite the proximity, there is almost no indication of the watery world just to the west—an unusual, often unappreciated, and now-rare landscape.

A tidal marsh is a remarkably flat, seemingly homogenous place, with elevational variation of just a few feet across several miles. Yet it is actually a diverse landscape. Microtopography—subtle variations in marsh elevation built by tides, currents, and plants—makes the difference between sinking waist-deep in mudflats and walking across the slightly higher marsh plain in sneakers. Walls of 10-foot-tall tule along brackish channels are juxtaposed against broad plains of low vegetation. Specialized plants dominate the environment; they're adapted to the combined challenges of submersion, salinity, and rapid change (the water levels fluctuate 5 to 6 feet every six hours). In this harsh landscape, most trees are excluded and flowers limited. But upon closer examination, expanses of pickleweed present a luminescent pattern of green, orange, and red. A native sunflower, Pacific gumplant, follows higher ground along tidal channels; a few willows and live oaks find small islands. Other hardy plants occupy distinct places along gradients of elevation and salinity.

The arms of the bay, with the winding bayous (called here "sloughs") in the swamps around them, were very beautiful, the effect heightened by the rugged mountains on the north and northwest.

—BOTANIST WILLIAM BREWER, 1861

Because of their transitional position between land and water, tidal marshlands are used by a wide range of animals and are among the most productive ecosystems on earth.[1] Tidal channels and young, low-elevation marshes provide foraging habitat for estuarine fish, including white sturgeon, Sacramento splittail, and juvenile steelhead and salmon. Migrating shorebirds feed in shallow marsh ponds and in the tidal flats along channels, while diving ducks use deeper ponds and sloughs.[2] The endangered salt marsh harvest mouse and California clapper rail, as well as rare plants such as Delta tule pea, soft bird's beak, and Mason's lilaeopsis, all depend on tidal marshes for their survival.

Lying at the bottom of the watershed, tidal marshlands also provide a reflection of upstream watershed processes: the characteristics of the tidal system inform us about the nature of the river and valley above. The historical prevalence of snags and floating trees in the Napa River's tidal channel shows that the river was able to recruit large woody debris from adjacent riparian forests and that fallen trees persisted in the channel. Similarly, the repeated dredging of gravel deposits from tidal waters indicates that the river was moving large amounts of gravel, maintaining and mobilizing riverine bars and spawning beds as part of downstream transport to the Bay.

Tidal marshlands and unvegetated tidal flats comprised most of the intertidal portion of San Francisco Bay. The much-noted reduction in the size of the Bay is primarily due to the loss of these habitats at the Bay's margins. Before diking and filling, nearly half (47%) of the Bay was exposed at low tide.[3] With the area of these shallow, intertidal habitats nearly equal to that of the deeper, subtidal waters, the Bay would expand and contract horizontally every six hours—almost doubling in size on a very high tide. Development has since reduced the subtidal waters by 10%. But the area of tidal marsh has decreased by 80%, so the Bay is now mostly (78%) open subtidal water. Because these broad, productive outer edges have been separated from the Bay by vertical walls (levees, seawalls, bulkheads), so that water now mostly just goes up and down, like in a bathtub, rather than in and out.

Along the lowest reaches of the Napa River, a similar local transformation took place but, in a dramatic shift, is now being reversed. In the 1850s, tidal marshlands covered 18,000 acres, representing one of San Francisco Bay's great tidal marsh complexes. In fact, the Napa marshes represented almost 10% of San Francisco Bay's tidal wetlands and about 3% of the coastal wetlands of the entire state of California.[4]

By the mid-20th-century, the river's marshlands had been eliminated. They were reclaimed, however, primarily by diking—the exclusion of tides by the construction of earthen levees around their margins—rather than by filling. As a result, the former marshlands remain mostly at or below sea level,

A tidal marsh pond near Slaughterhouse Point, Vallejo. *Photograph April 1, 2010.*

The land from Soscol down toward the bay was covered with tules, and in the dry summer and fall the elk would come down from the hills and browse in the marshes.

—IGNACIO VALLEJO,
RECORDED BY GUY WINFREY (1953)

protected only by narrow berms vulnerable to wave erosion, earthquakes, flood damage, and rising sea levels. In the past decade, the California State Coastal Conservancy, California Department of Fish and Game, U.S. Fish and Wildlife Service, Napa County Flood Control and Water Conservation District, and the city of American Canyon have acquired over 10,000 acres for restoration. These combined efforts have initiated a new phase in the management of the marshlands. Because of its size and restoration potential, the current redesign of the Napa marshlands is of regional and statewide significance.

But redesigning this landscape is nothing new. Occupying the lowest part of Napa Valley at its interface with the Bay, the tidal marshlands lie at the edge of human control. Their stories particularly illustrate the back-and-forth dynamics of the natural and cultural processes shaping landscapes: a continual calibration between human activity and physical forces. Over the course of just a handful of generations, newcomers to the valley have collectively imposed a series of dramatic modifications, experienced the system's response, and, as information and priorities change, adjusted goals and approaches.

Through indigenous, Spanish, Mexican, and early American history, people have recognized that tidal marshlands provide useful products and services, including salt from natural salt-producing ponds, deepwater channels for navigation, waterfowl hunting grounds, and protection from flooding. As the tidal system was modified to increase some of these functions, others diminished. The introduction of new land uses has often been temporary and had unintended consequences, such as the proliferation of mosquito habitat and the downsizing of tidal channels in response to reduced tidal area.[5] Levees and other infrastructure require continual and costly maintenance. The resulting contemporary landscape is a snapshot of human adaptation to the local landscape, and the landscape's response to those changes.

To study this ever-shifting landscape, in this chapter we explore the characteristics of the historical tidal marshlands, their history of dredging and reclamation, and contemporary processes of restoration and sea level rise.

Napa River's tidal marshlands, circa 1840. The Tulucay rancho *diseño* captures the basic physical geography of the marsh and surrounding landscape. *Estero de Napa* is tidal Napa River; adjacent blue areas are tidal marsh *(Tular)*. Hills extend all the way to the river at the Suscol landing *(Embarcadero)*. The dashed brown line on the left, labeled *Camino,* corresponds to the present-day Napa-Vallejo Highway (Hwy. 29-Soscol Ave.). *USDC circa 1840c, courtesy of The Bancroft Library, UC Berkeley.*

Pickleweed close-up. Photograph by Susan Schwartzenberg.

KERR'S MAP

Hidden within the subtle marsh topography lies a highly organized and diverse landscape. David Kerr's 1858 U.S. Coast Survey topographic map ("T-sheet") remains one of the best illustrations of the intricate patterns that emerge in a large natural tidal marsh.

In the map, details of which are shown on the opposite page, the vegetated marsh plains are indicated by closely spaced horizontal lines **ⓐ**.[6] Sinuous tidal channels or sloughs **ⓑ**, ranging from a few feet to a half mile wide, carry tides into and out of the marshland. (Comparison to later aerial photographs and other parts of the Bay suggests that Kerr mapped most of the channels precisely, except for the smallest, single-line channels, which were depicted more diagrammatically.) The dotted line **ⓒ** shows the extent of open water at low tide (that is, the minimum amount of water in the channel) and marks the edge of tidal flats along the larger channels: the flats are exposed at low tide and flooded when the tide rises. If the tide is very high, it spreads across the marsh plain (covering the entire lined area).[7]

Away from the channels, scattered marsh ponds (ranging in size from 1,000 square feet to over 40 acres) are shown as enclosed white shapes **ⓓ**. Larger ponds often paralleled the landward edge of the marsh; these were flooded at the highest tides and often evaporated by late summer into dry, playa-like pannes with salt-encrusted surfaces. Higher ground is shown with "tufts" indicating grassland or pasture **ⓔ**. Red contour lines indicate higher ground on islands and adjacent hills **ⓕ**. A few trees are recorded as small, clover-like shapes **ⓖ**.

Because of the small size of existing marshes, we have few examples of these landscape-scale patterns. Today Kerr's map, with its extreme level of detail, provides a link between the marsh's past and its future.

Nine details of Kerr's T-sheet at its original 1:10,000 scale. (opposite page) Each detail covers 115 acres. The full map is over 7 feet long, extending from American Canyon to the city of Napa. A larger area is shown on page 133.

Top row (left to right) • A sinuous channel network spreads through the marsh. • Ponds, marsh plain, and channels surround a grassy island (still called Green Island today); a few trees can be seen on the island's northern edge. • Small channels branch from a large slough; dotted lines show the area of water at low tide.

Middle row • A creek with scattered large riparian trees empties into the marsh; red lines show the 25-foot contour. • A chain of marsh ponds lies opposite the Napa River's Horseshoe Bend. • Groups of small ponds and large oaks follow the complex margin between land and marsh.

Bottom row • The Napa River channel abuts dry land, making a natural landing site at Suscol. • An oddly shaped tidal channel perhaps incorporates recently captured ponds. • Tidal channel, ponds, and early agriculture (shown as diagonal solid and dotted lines) lie at the marsh edge.

TIDAL MARSHLANDS • 127

THE TULE LANDS

Tule swamps, forming at first narrow belts…gradually widen as you descend the valley, until, at its mouth, they usurp nearly the whole of its surface.
—AUTHOR BAYARD TAYLOR, 1862

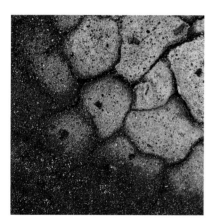

To enhance our understanding of the marshes shown by Kerr's map, we turn to 19th-century descriptions. These accounts consistently referred to tules, indicating tall bulrushes of the sedge family associated with fresh and brackish waters. Mexican *diseños* used the term *tular* ("place of tules") to label depictions of the marshlands; American accounts described the "tule lands."[8] James Clyman noted "greate *[sic]* and Extensive Bull Rush marches *[sic]*" in 1845.[9] William Brewer described the marshlands from his campsite at Suscol in 1861, confirming the vigorous, tall marsh vegetation: "The swamps bordering all the rivers, bays, or lakes, are covered with a tall rush, 10 or 12 feet high, called 'tule,' which dries up where it joins arable land."[10]

The term "tule" can be used imprecisely to refer to wetland vegetation of any kind, but its consistent use in references to the tidal marshlands along the Napa River distinguishes these marshlands from more-saline wetlands of the central and south San Francisco Bay.[11] There is, in fact, substantial evidence suggesting the Napa marshes were relatively fresh before diking and water diversions. One source of information is accounts of irrigation using the tidal waters. In 1885, local farmer John A. Stanly (of Stanly Lane, near the Southern Crossing) reported: "A local advantage, incident to my land, is the facility of irrigating it in June and July, when the tide water is fresh, thereby giving a new growth of grass when the grass on uplands is dry."[12]

Recollecting conditions in the first half of the 20th century, Walter Carvelli described similar practices further upstream, in the vicinity of present-day JFK Memorial Park: "The river there was pretty fresh most all year long, least during some years. There was a ranch across the river from there up just a ways called Rattle Stock. They used to get water from the river for crops."[13]

The dominance of tules in textual accounts does not mean, however, that the marshlands consisted of only a single species. Brewer was describing their general appearance from the upland margin; another river traveler hinted at more heterogeneity: "on both sides of the river a tract of low, level land (in some places covered with a heavy growth of tule)."[14] Historical evidence suggests that, away from the channels, the marsh plains were characterized by lower vegetation. Here drainage was less effective and salts accumulated in the peat soils, resulting in the low, salt-tolerant "salt grass" and "turf" that Stanly described on his lands prior to reclamation. Patterns of soil salinity created a complex vegetation mosaic, suggested in Carpenter and Cosby's description of "a growth of tules, sedges, salt-tolerant grasses, and other plants."[15]

These vegetation patterns correspond substantially to historical accounts of tidal marshlands in Suisun and the western Delta.[16] The presence of species such as Mason's lilaeopsis, Delta tule pea, and Sacramento splittail[17] also affirms the affinity between the Napa marshes and the larger fresh and brackish marshlands of the greater San Francisco estuary. The Napa River marshes constituted a smaller, secondary estuary within the system, what Carpenter and Cosby called the "delta land at the mouth of Napa River."[18]

There is also some suggestion that the Napa River received significant freshwater directly from Delta outflows. Summer flows coming through the Delta may have been naturally pushed upstream at Mare Island by the tides, making the tidal Napa River and its marshlands fresher than they otherwise would have been—and allowing Napa farmers to irrigate with tidewater into the summer. Carvelli stated, "when we started to lose the Sacramento River water everything got a lot saltier…we lost that big river water."[19]

Cows grazing in Thompson's reclaimed brackish marshes at the Suscol landing. (top) *Smith and Elliott 1878.*

Contemporary Napa River. (bottom) Narrow zones of tidal flat and tules can be seen along the channel in this view just upstream of Stanly Lane, marked by the row of eucalyptus in the background. *Photograph November 27, 2009.*

The land drainage alone might afford a fitful and uncertain navigation for a part of the year, but it is the daily influence of the tide which gives the river permanent value for navigation.

—COL. G. H. MENDELL, 1885

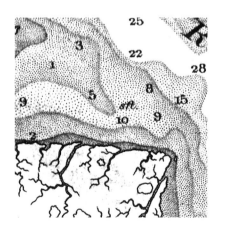

Soundings along the Napa River, 1860. Numbers indicate the depth in feet below mean lower low water. *Alden 1860, courtesy of NOAA.*

DREDGING THE TIDAL RIVER

The Napa River's watershed was too small and flows too seasonally variable to produce a commercially navigable river from rainfall alone. Instead, the tides made waterborne transport possible, initiating American development of the valley. Without modification, the tidal reach of the Napa River supported agricultural commerce for over four decades, beginning in the 1850s. Dredging by the U.S. Army Corps of Engineers was not essential for navigation, but was initiated to allow bigger ships and unimpeded travel at low tide. Corps dredging took place within the tidal reaches downstream of the town of Napa, beginning in the late 19th century.

While most of the river up to the town of Napa was 5 feet or deeper at low water in the mid-1800s, the channel was interspersed with shallow shoals or bars, some less than 1 foot deep.[20] These low tide obstructions forced the arrival and departure of steamers and other vessels to be timed "when the phase of the tide will permit."[21] This timing of course varied daily, making scheduling complicated and necessitating the frequent use of landings 4 miles downstream from Napa at Suscol.[22] A fast-moving bend could be 20 feet deep while an adjacent reach was just 3 feet deep at low water—impassable by a steamer with a 4-foot draft.[23] This heterogeneity of the pre-dredging tidal channel, while impeding navigation, created a range of depths, substrates, and flow velocities that provided areas for spawning, feeding, high-flow refuge, and other needs of native fishes.

In 1885, Army Corps engineer George Mendell recommended increasing the minimum depth to a consistent 4 feet, but in practice the federal improvement project achieved 6 feet, making the river reliably navigable for boats drawing 4 to 5 feet of water. Excavated material (an estimated 145,000 cubic yards) was deposited on the adjacent banks—the "easiest and best place."[24] In 1914, the Corps also recommended straightening the river to remove "several very bad bends."[25] In 1935, the project was modified to 8 feet deep, and dredging continues to date (authorized depths are 10–15 feet).[26] As a result of these continuous efforts, along with the concurrent diking of adjacent marshlands, the winding, heterogeneous tidal river of steamboat times became a straighter, wider, and deeper channel, enclosed by levees on both sides.

Some channel complexity has recently been reestablished, however, as part of the Napa River Flood Protection Project, which has moved levees away from the channel along several miles of the tidal river. Tidal flats, marshes, and adjacent floodplain have been created through excavation of a broader channel with shallow shoals and benches. Interestingly, this is an area that historically had quite different characteristics, due to the greater dominance of riverine (rather than tidal) processes: a relatively narrow channel with scattered riparian trees on its natural levees. In effect, desirable historical characteristics of the lower tidal river (tidal flats, broad channel) have been moved closer to town to gain their natural flood protection functions.

"Government Dredge at Work Deepening Napa River." Looking south from the 3rd Street Bridge, local farmer and photographer H. A. Darms took this image of a clamshell dredger. It was widely reproduced as a postcard, and also published (with the above title) in his 1908 photographic atlas. Piles of dredged material can be seen deposited on the left bank. *Darms 1908.*

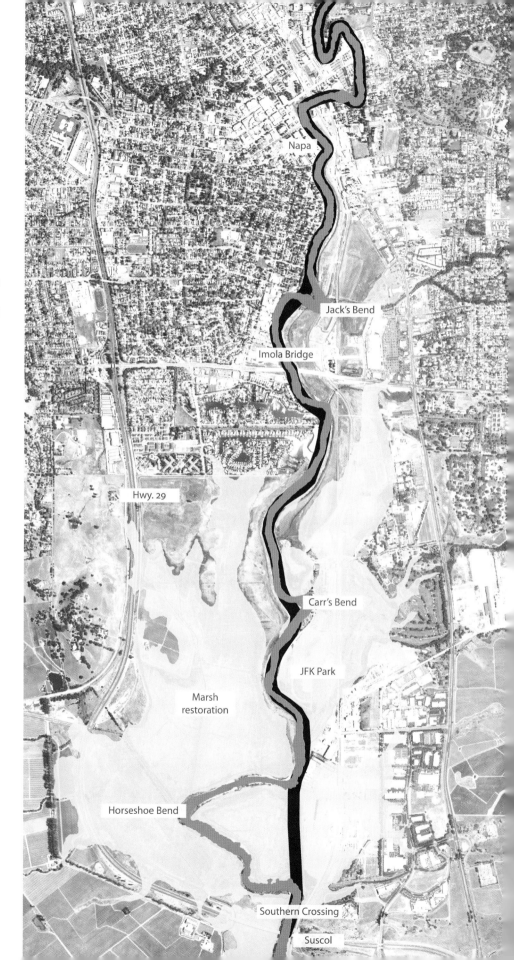

Widening and straightening of the Napa River's tidal reach. Federal dredging projects since 1888 have substantially straightened the tidal channel that serves as the approach to Napa's port. The river's width has also been expanded. The historical channel (blue) from David Kerr's 1858 T-sheet (opposite page) is overlaid on the contemporary channel (red). The green area shows historical tidal marsh extent, all of which was reclaimed by the 1940s. *Kerr 1858, courtesy of NOAA.*

Proposed river cutoffs, 1914. (inset, opposite page) The improvements proposed in this Army Corps map and similar projects reduced the distance by boat from the Southern Crossing (Butler Bridge on Highway 29) to downtown Napa by more than a mile (from 5.4 to 4.3 miles; see page 94). As a result, the channel's sinuosity in this reach (a measure of its "curviness") decreased from 1.4 to 1.1. While the river became a more efficient transportation corridor, there was an associated reduction in channel complexity and habitat diversity. *Rees 1914.*

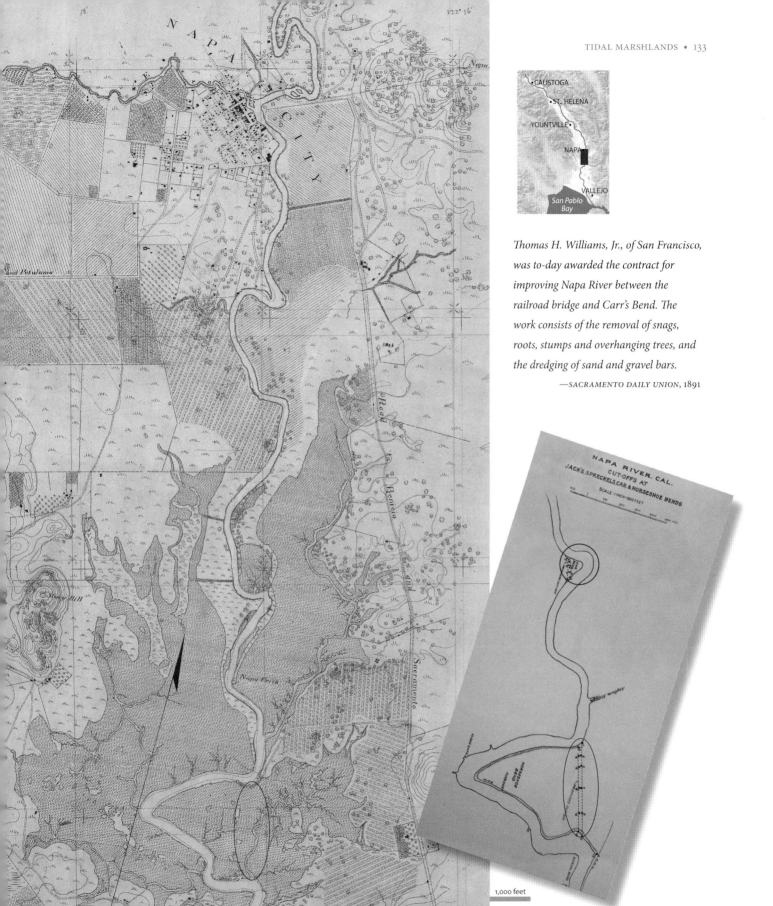

TIDAL MARSHLANDS • 133

Thomas H. Williams, Jr., of San Francisco, was to-day awarded the contract for improving Napa River between the railroad bridge and Carr's Bend. The work consists of the removal of snags, roots, stumps and overhanging trees, and the dredging of sand and gravel bars.
—SACRAMENTO DAILY UNION, 1891

RECLAMATION AND RESTORATION

Years ago bass were so numerous in the lower reaches of Napa's sloughs that a man rowing a boat would strike a fish every few minutes with his oars. In recent years bass fishing in these sloughs has been largely abandoned because almost every slough that formerly afforded good fishing has been leveed off.

—NAPA RIVER ANGLER W. P. WEST, CIRCA 1926 [29]

Napa's tidal marshlands and people have coexisted since the marshlands' formation. When the marshes formed, as the rate of sea level rise slowed several thousand years ago, native peoples already occupied the margins of the expanding Bay. For generations, native tribes used the marshes for catching fish, hunting waterfowl, collecting salt, and transportation; 19th-century Mexican and American cultures used the marshes similarly. Priorities shifted in the 1870s as farmers began to alter tidal hydrology to favor agriculture. Local author Menefee announced "the reclamation of a large body of tule land" taking place at Thompson's farm near Suscol in 1873. Five years later, the successful site was described as "meadow land reclaimed from…the overflowed tidelands of the Napa River."[27]

Yet it took time for reclamation to take hold. Many early levees failed. An 1885 article reported that "there have been attempts to levee or dike nearly all the marsh land…but the levees were not built strong enough to resist the pressure of the tide."[28] By 1914, however, USGS topographic maps showed levees constructed around the majority of the Napa River's marshlands. Robert Byrns, leading the resurvey of the marshlands for the U.S. Coast and Geodetic Survey's updated T-sheet in 1922, summarized both the transformation and its ordinariness, as similar changes were taking place throughout San Francisco Bay:

From marsh to pond to marsh. Salt ponds encircle Green Island, which was formerly surrounded by tidal marshlands (see page 127). These ponds are now being restored to marshland. *Photograph April 11, 2009.*

No extraordinary changes were found. The most noticeable were the levees that have been built since the last survey and the amount of reclaimed land resulting from them. Practically all of this reclaimed land is under cultivation or being used for grazing purposes. The levees are good, substantial ones and more of them are being built at the present time. As noted on the sheet, a number of small lakes and sloughs have ceased to exist due to the levees.[30]

By 1937, all the major historical marshlands along the Napa River (except, for unknown reasons, Fagan Marsh) had been diked.[31] Reclaimed marshes were managed for crops but also for "duck shooting clubs" and, beginning in the 1950s, salt production.[32] One unintended consequence of reclamation was the proliferation of mosquito habitat, leading to extensive control efforts.[33]

In 1978, 120 years after David Kerr produced his map, 18,000 acres of tidal marshland had diminished to about 2,900 acres, an 84% decline. Since then the trajectory has reversed: by 2009 there were 4,300 acres. In addition, 4,000 acres of diked baylands have recently been reopened to the tides and are anticipated to evolve into vegetated marshes. In the next few years, another 3,400 acres are planned for restoration. If successful, these projects will represent a remarkable recovery of the Napa River's tidal marshes: an increase of 350% since the 1970s, to about 65% of the historical marshland extent.

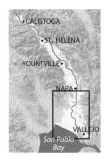

Tidal marsh trajectory, 1858–2009. Tidal marshlands along the Napa River were almost completely eliminated during the first few decades of the 20th century. For over a half century, few marshes existed but recently their area has increased.[34]

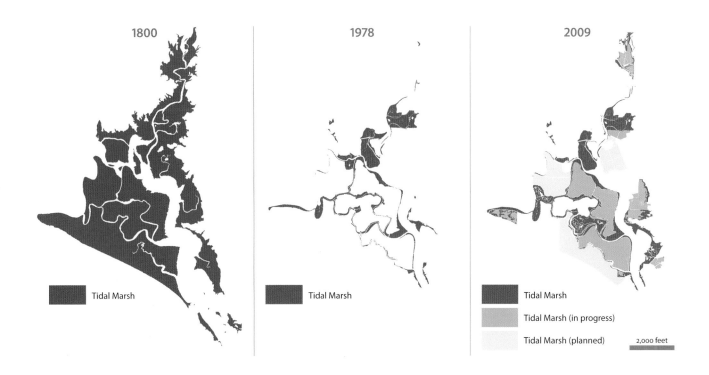

MARSH CHRONOLOGIES

Tidal marshland returns. (below, and continuing on opposite page) In 1858, surveyor David Kerr carefully traced the boundary between tidal marsh **a** and pasture land **b** near Horseshoe Bend. By 1942, the former outline—which is shown with a green line—was unrecognizable, as the area had been diked and converted to agriculture. Since restoration in 2001 as part of the Napa River Flood Protection Project, the tides have precisely reoccupied their former extent up to Highway 29 **c** and the Napa Yacht Club development **d**. In the 2009 image, new marsh **e** and mudflats **f** are visible. Unvegetated areas appear to be forming in the same locations as the large marshy ponds evident in 1858 **g**.

The figures on this and the following spreads illustrate change at three places. The marshlands known as the South Wetlands Opportunity Area were reopened to the tides in 2001 as part of the Napa River Flood Protection Project. The former salt pond known as Pond 5 was breached as part of the Napa River Salt Marsh Restoration Project in 2006. The spread on pages 138–139 examines Fagan Marsh—the largest never-diked tract of historical marshland along the river—and neighboring Bull Island.

A small, hundred-acre marsh island surrounded by branching channels of the Napa River, Bull Island demonstrates a complex sequence of human modification and response by the natural system. Cornelius Eiling Davis, the great-grandfather of local resident Lydia M. Money, completed diking the island in the 1870s. The family used the island's rich peat soils to grow winter rye and wheat for nearby cattle ranches for several decades. The land then remained fallow for much of the first half of the 20th century (~1920–1949), but the levees held. The island was cultivated again (for Red Pontiac seed potatoes) for an intensive six-year period until floods in 1955 and 1958 breached the perimeter levee. Salt deposits rapidly accumulated on the land surface as the semi-enclosed area unintentionally began to function like a shallow salt-evaporating pond. The salt accumulation and expense of levee repair precluded subsequent return to agriculture; the land gradually reverted to marsh.[35] In 1997 the island was acquired by the Land Trust of Napa County for transfer to the state as a public reserve.

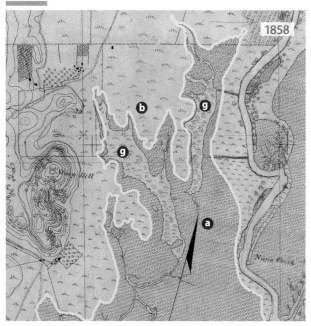

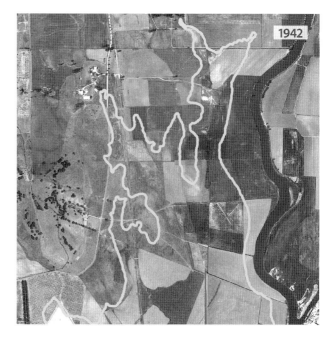

TIDAL MARSHLANDS • 137

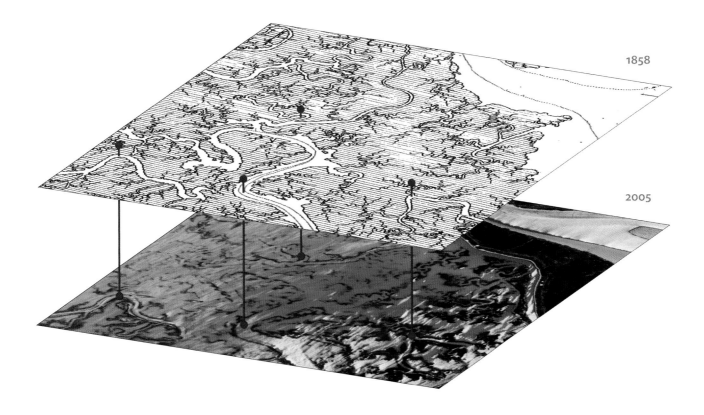

The marsh at Pond 5. (above) Historical tidal marsh channels were still visible in 2005 imagery of this salt pond, even though they had been disconnected from the river for a century or more. This area is now being restored by federal, state, and local agencies. Some levee breaches are being located to reoccupy the "ghost" channels so that they can guide the reestablished tidal flows.

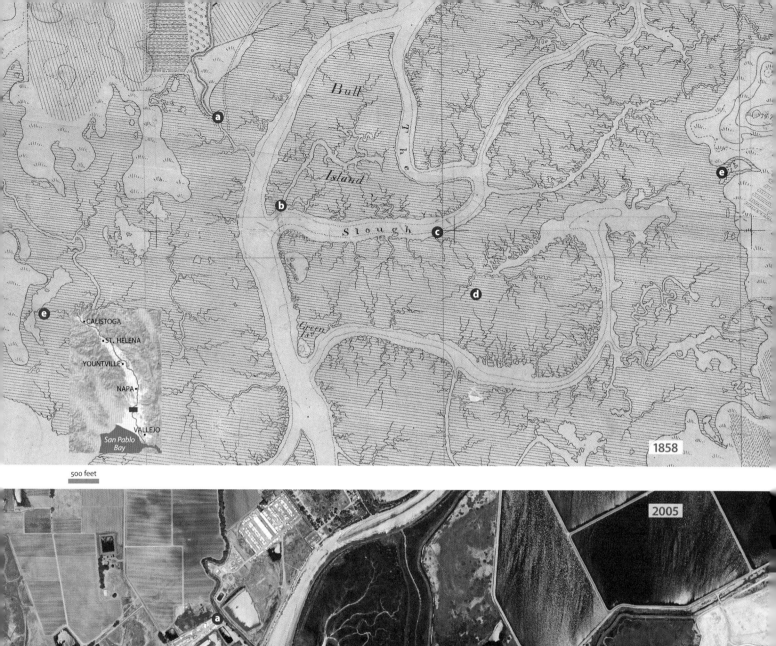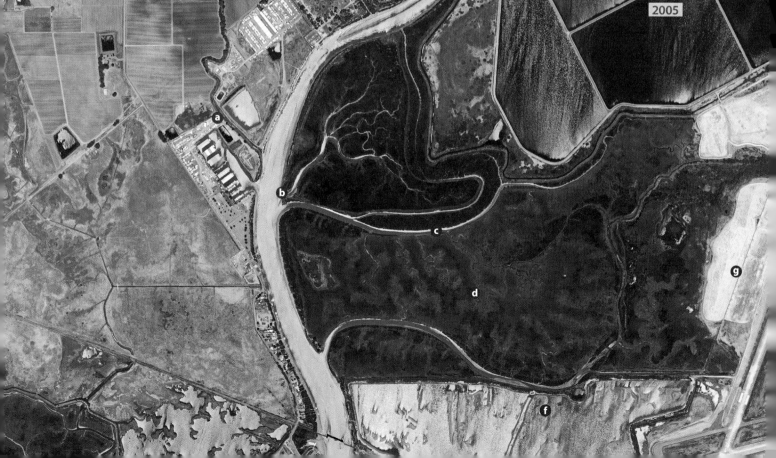

BULL ISLAND AND FAGAN MARSH ECOLOGICAL RESERVE

In the 1850s, roads led to the natural landing points where tidal channels came close to land. Napa Valley Marina expanded upon one of these sites, occupying the former marshland adjacent to the creek **a**. Dredging is currently required to keep the marina open.

In 1914, Bull Island's main entry channel was visible but inactive, cut off by the surrounding levee **b**. In the contemporary view, the channel has reopened and extended to the north, now branching off the main channel; the perimeter levee can still be faintly seen.

As branches of the tidal channel network have been cut off, the larger trunk channels have downsized **c**. The longer, eastern route around Bull Island was actually the deeper channel, referred to as "Steamboat Slough," or "The Slough." With the disconnection of the sloughs and marshland to the east, the channel filled in with sediment and is no longer navigable by larger boats. Fagan Marsh **d** is the largest area of the Napa River marshland never diked. Fagan Marsh retains many of its historical characteristics, including extensive tidal channels and a small island on its western margin. But the entry channel, which supported large mudflats, has filled in dramatically.

Some of the tidal marsh pannes at the landward edge **e** likely created evaporative salt deposits in the late summer. Expanding upon this natural process, human industry converted much of the marshland by the lower river to large-scale saltwater evaporation, with the concomitant effect of reducing tidal scour and flood storage capacity. The salt ponds to the south **f** were reopened to the tides in 2009.

The eastern edge of the Fagan Marsh near the Napa airport runway **g** appears natural, but actually reflects several hundred yards of artificial fill. However, this area may be reoccupied by the tides as sea level rises, providing an important transitional zone.

Napa marshes through time. On the opposite page, the marshes in 1858 (top) and 2005 (bottom). Below, an oblique view of the 1914 USGS quadrangle shows an intermediate and slightly broader view. The lines of brown hatchmarks **h** show the newly constructed levees along the Napa River, separating the channel from most of its adjacent marshlands. Fagan Marsh lies between Bull Island and the railroad **d**. *Below: USGS 1916. Opposite page, top: Kerr 1858, courtesy of NOAA.*

THE EDGE

The most easily overlooked component of tidal marshland—but perhaps most important—is its landward edge. This transitional zone between tideland and dryland contributes an array of ecological functions. Diverse plants occupy this margin, including several now-rare species.[36] Many animals regularly cross the land-marsh ecotone, coming to the marsh to feed or, in the case of the salt marsh harvest mouse, retreating to land when the marsh is flooded during extreme high tides.[37] These areas are typically the first parts of the marsh to be impacted by human activity and are among the most underrepresented elements of the historical landscape today.

The landward edge also plays a critical role in the future of a tidal marsh. The seas have been rising naturally since the last ice age, forcing tidal marshes to adjust to new water levels or be submerged. If plant growth and sediment supply are sufficient, the marsh can keep pace through vertical growth. (Thus the long-term fate of the Napa River's tidal marshlands is in large part dependent on the delivery of sediment from the watershed to the Bay margin.) Tidal marshes can also persist during times of rapid sea level rise through gradual migration inland across adjacent lowlands, a process called estuarine transgression. Around most of San Francisco Bay, however, there is little room between the rising Bay and adjacent development. With no place to go and accelerated sea level rise due to climate change, marshes may drown.

Some areas around the edge of Napa's marshlands become particularly important in this context. The Napa Valley has a few of the relatively rare non-urbanized Bay margins—lowlands immediately inland from existing or potential marsh. These areas should be recognized for their potential to provide key ecological services, from the restoration of ecotone communities to the maintenance of tidal marshes as sea level rises.[38]

There were literally millions of ducks and other waterfowl inhabiting the ten or so miles of marsh, clear to San Pablo Bay, many of them nesting in the grassy upland.

—STEWART DUHIG
DESCRIBING HIS FATHER'S RECOLLECTIONS OF THE MARSHLANDS CIRCA 1870[39]

Plant diversity at tidal marsh ecotone. On his train ride along the American Canyon marshes, Robert Louis Stevenson captured the unusual, transitional character of the margin between marsh and land. A dominant feature historically, such gradual transitions are rare now. "In long, straggling, gleaming arms, the Bay died out among the grass; there were few trees and few enclosures; the sun shone wide over open uplands."[40] *Photograph by Susan Schwartzenberg, April 18, 2009.*

Rising seas and the marsh edge. Current projections of 55 inches in sea level rise by 2100[41] would cover the area in blue. Comparison with the green outline (historical tidal marsh extent) shows that, in the absence of significant new infrastructure, the Bay will reoccupy nearly all of its former area, and a little more. Depending on rates of erosion, sediment supply, and other factors, some marshes may be submerged or not able to reestablish in their former locations. Compared to much of the San Francisco Bay, however, the landward margin along the Napa River tidal marshlands is not intensively urbanized, offering the potential for inland migration of the marsh and upland ecotone.

 Sea level rise by 2100

Historical tidal marsh extent

2,000 feet

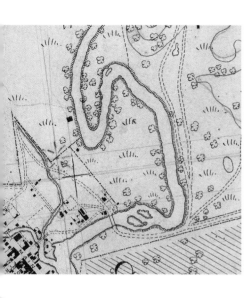

7 · LANDSCAPE TRANSFORMATION AND RESILIENCE

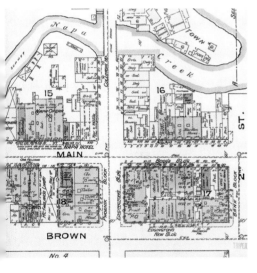

As the previous chapters illustrate, the Napa Valley landscape tells a complex story. In 200 years—just three long lifetimes—tidal marshes have been drained and turned into potato fields, flooded as salt ponds, and, in a number of cases, returned to marsh again. Some streams have been removed while others have been extended. Oak savannas have become cattle ranches, fields of wheat, prune orchards, vineyards, and cities. The transformation of the past two centuries has created a dynamic and productive landscape revered by people throughout the world, and also had unintended consequences.

In recent years, there has been increased recognition of these continuing impacts: the loss of native species, the decline of river health, and the effects of landscape modifications on land erosion, flood risk, stream flow, and other valued landscape functions. Entirely new programs and projects have been created in response, initiating a new phase in the management of the valley. A significant portion of the Napa River's tidal marshlands is being restored to provide flood capacity, a buffer against sea level rise, and habitat for fish and wildlife. Buildings and infrastructure in downtown Napa have been moved to provide more space for the river. Farmers have initiated innovative efforts such as the Napa Sustainable Winegrowing Group and the Napa River Rutherford Reach Restoration Project.

Because of the success and preservation of agriculture, urban and suburban development does not yet dominate the Napa Valley, as it does much of the Bay Area. This trajectory of development could still occur if the agricultural preserve is not maintained and reauthorized over time. But it is also possible that, because of its particular cultural and physical attributes, the Napa Valley could continue to sustain agriculture while creating a resilient landscape supporting enhanced ecological functions and processes. The Napa Valley may be particularly suited to draw upon its natural heritage in creative ways.

More so than in many places, the land is still appreciated in Napa. The subtleties of alluvial fans and weather patterns are studied and discussed, respected as the foundation for the local culture and economy. Within the agricultural community there is tremendous knowledge of the topography, soils, and microclimates of the valley, and their expression in the subtle characteristics of local wines. The idea of *terroir*—the influence of the land on the nature of a wine—may be the fundamental concept underlying today's Napa Valley, expressed in the quality of its wines, the diversity of appellations, and the economic value associated with the Napa name. In Napa, the sense of place is still celebrated.

At the same time, the Napa Valley has become an iconic landscape, a powerful international symbol. Orderly vineyards and grand estates convey an image of a harmonious, pastoral landscape. Yet the landscape patterns we see today are not placid or stable, or even particularly old. From the orchards of 50 years ago to the wheat fields and oak savannas of 150 years ago, the landscape is still being shaped. Compared to the famous and ancient wine regions of Europe, Napa Valley culture is still young, still in the process of establishing its relationship with the natural landscape and proving its sustainability. As a result, native species and habitats are not long gone. Old oaks in the corners of vineyards reach back in time to a landscape managed by the Caymus and Canijolmano, a newness unthinkable in France.

Part of the Napa mystique is in fact the rawness and vibrancy of the land, wildness mixed with refinement—rugged hills and remnant valley oaks framing vineyards in wine labels. It is interesting to consider what role the sense of nature's bounty plays in the Napa image. Salmon, oaks, songbirds, and all the other elements of the historical Napa Valley are also integral parts of the landscape itself, the *terroir* of the region. The conception of Napa as a place, and the way its underlying landscape is treated, will determine not only the future character of the valley, but its role in the world's imagination. By enhancing and sustaining the local *terroir*, Napa, unlike almost anywhere else, can provide an example to the rest of the world of beauty and productivity, ecology and commerce.

USING HISTORICAL ECOLOGY

By creating a longer-term understanding of the landscape, historical ecology helps environmental managers, scientists, and local residents develop strategies to enhance the ecological health of a region. Information about the historical landscape improves our knowledge of the relationship between governing physical processes, native habitats, and the ecological functions they provided. Rather than providing a literal template for re-creating past landscapes, historical ecology improves our understanding of the processes that created and maintained the habitats and functions we seek to restore. This landscape-level perspective facilitates accurate identification of the missing elements in our contemporary landscape and shows what we might recover within the context of current and projected future conditions.[1] In the absence of historical ecology investigations, assumptions about historical conditions have often been found to be inaccurate, leading to misguided or unsuccessful investments in environmental restoration.[2]

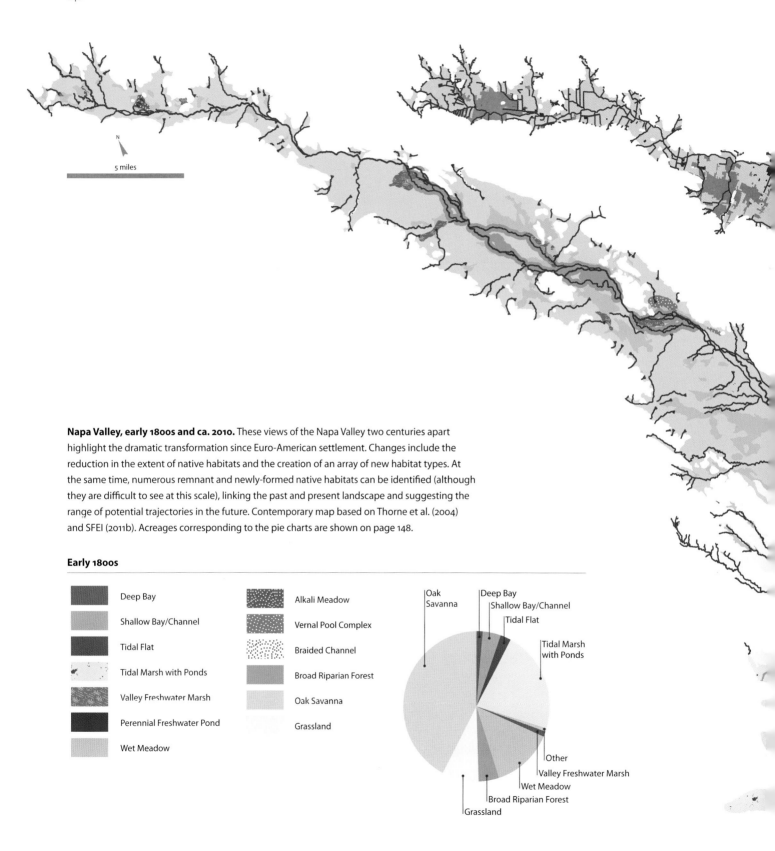

Napa Valley, early 1800s and ca. 2010. These views of the Napa Valley two centuries apart highlight the dramatic transformation since Euro-American settlement. Changes include the reduction in the extent of native habitats and the creation of an array of new habitat types. At the same time, numerous remnant and newly-formed native habitats can be identified (although they are difficult to see at this scale), linking the past and present landscape and suggesting the range of potential trajectories in the future. Contemporary map based on Thorne et al. (2004) and SFEI (2011b). Acreages corresponding to the pie charts are shown on page 148.

Early 1800s

- Deep Bay
- Shallow Bay/Channel
- Tidal Flat
- Tidal Marsh with Ponds
- Valley Freshwater Marsh
- Perennial Freshwater Pond
- Wet Meadow
- Alkali Meadow
- Vernal Pool Complex
- Braided Channel
- Broad Riparian Forest
- Oak Savanna
- Grassland

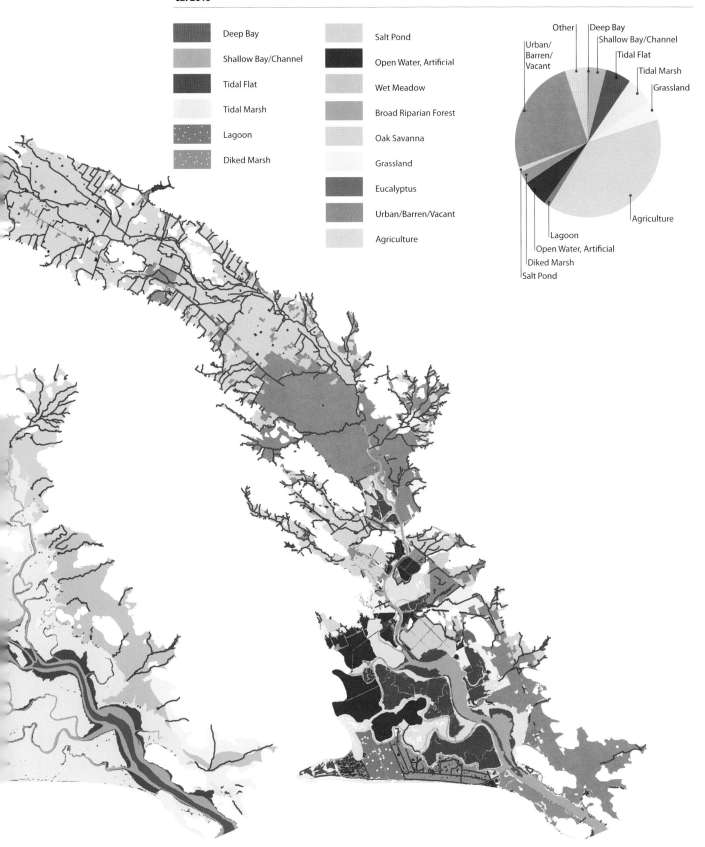

A historical landscape perspective, when combined with contemporary fieldwork, allows a better understanding of the contemporary and potential future landscape. First, through the comparison of past and present information, we recognize historical habitat remnants, elements that have persisted intact despite surrounding modifications. Analyzing what we call the "remnicity" of the landscape reveals certain places that deserve our attention and appreciation. Fagan Marsh, the old valley oaks lining Highway 29, and the small remnant freshwater wetlands demonstrate the viability of native habitats within the developed landscape. Such sites can serve as nodes for restoration. They provide a glimpse of the historical landscape and ideas for the future.

Secondly, historical perspective reveals the responsiveness of the landscape. People have changed the land, but the land continues to respond, adjusting to new conditions. The oak orchard, the returning beavers, the new riparian forest on redirected Sulphur Creek, and other "accidental restorations" show that the processes that created the Napa Valley of the early 1800s are more resilient than we might imagine. In a very tangible sense, the Napa Valley is still the Napa Valley. The basic patterns of soil and topography are still fundamental controls; former wetland areas are still wet. Steelhead still return each year. Scrub jays continue to plant acorns, which are direct descendants of generations of valley oaks. The forces of rain, sun, tides, and ecology continue to operate on the thin layer of human construction that is our roads, buildings, pipes, and fields. That layer is constantly changing, being rebuilt and redesigned as crops are replanted,

Extent of major habitat types of the Napa Valley, early 1800s and ca. 2010. Most of the native habitats of Napa Valley have declined precipitously, but different trends can be identified. The overall reduction of tidal marsh has been less extreme due to restoration in the past decade. The extent of tidal flat has actually increased, as former salt ponds have been reopened to the tides in the process of restoring tidal marsh. Most of the deep bay has become shallow bay, which, as a result, has experienced little net change. The most extensive new features on the landscape are agriculture, urban areas, and artificial open waters (including reservoirs and wastewater treatment ponds). Valley Wetlands includes Valley Freshwater Marsh (829 acres), Wet Meadow (11,751), Alkali Meadow (116), Vernal Pool/Swale Area (287), and Perennial Pond (13). Contemporary data based on Thorne et al. (2004) and SFEI (2011b). Areas are in acres.

Habitat Type	Early 1800s	Circa 2010	% Change
Deep Bay/Channel	1,147	161	−86%
Shallow Bay/Channel	3,385	3,298	−3%
Tidal Flat	1,971	4,836	145%
Tidal Marsh	18,726	4,841	−74%
Valley Wetland	12,996	105	−99%
Broad Riparian Forest	3,757	53	−99%
Grassland/Wildflower Field	6,894	3,694	−46%
Oak Savanna	35,811	114	−100%
Agriculture	–	32,787	–
Lagoon	–	1,076	–
Open Water, Artificial	–	4,933	–
Diked Marsh	–	2,296	–
Salt Pond	–	892	–
Urban/Barren/Vacant	–	21,481	–
Other	155	4,275	–
Total	**84,842 acres**	**84,842 acres**	–

buildings replaced, bridges and levees reconstructed. History helps us to see the living landscape around us.

Most fundamentally, historical ecology helps define choices: both the ones made in the past, and the ones that remain to be made. The Napa Valley, and much of California, embodies a contradiction and opportunity. The extreme transformation of the past few generations renders the recent historical landscape largely distant and unknown. But the valley of 50 years from now, could, if we wanted, include much of what was lost. It remains our choice how much we want the native species, habitats, and functions of the Napa Valley to be part of our future.

As we recognize the existing potential in the landscape, we can see more clearly the choices of today. From a longer-term perspective, our society has inhabited this valley for a very brief time, during an unusually stable climatic period and with relatively short-term priorities. As we get to know the historical landscape, we realize that many of its functions are compatible and even desirable within our developed landscape. We still want highways and electricity, safety from floods, a thriving agriculture and economy. But we also want a landscape resilient to climate change, a reliable water supply, a healthy river, fish and wildlife for our children.

RESILIENT LANDSCAPES

As we plan for a more variable climate in the future, we can examine the characteristics of the historical landscape that provided resilience and flexibility in response to dynamic conditions. The historical Napa Valley was well adapted to a highly variable climatic regime, providing diverse ecological functions while buffering the effects of environmental extremes. An array of characteristics helped modulate the effects of drought, flood, and variations in annual rainfall. Alluvial fans, spreading streams, and the river's floodplain dissipated flood energy while recharging groundwater. Valley wetlands, beaver ponds, side channels, and fallen trees all helped retain surface water through the dry season, preserving essential resources for plants and wildlife. Long-lived valley oaks were uniquely adapted to provide reliable shade and habitat across the valley floor through centuries of climatic variation.

One of the important characteristics of the historical landscape was these refugia: reliable habitats that helped populations of native species survive extreme conditions.[3] In the historical Napa Valley, native fish could retreat to the safety of side channels during floods. In times of drought, fish, amphibians, and other wildlife could rely on a few perennial wetlands and stream reaches located in the most favorable topographic and hydrological settings. While most water sources dried up, these would stay wet,

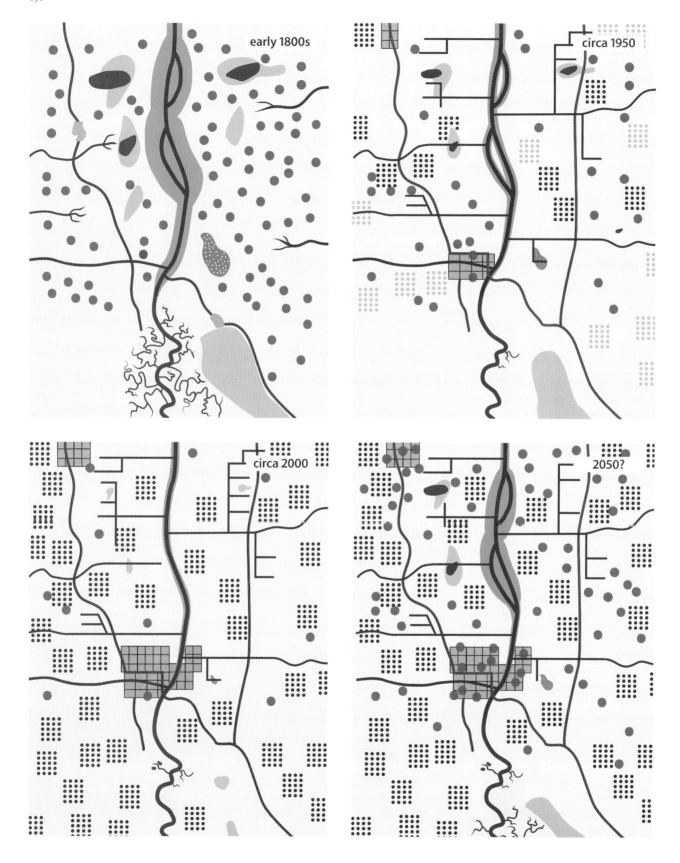

allowing small populations to survive difficult times. The reduction and simplification of habitats has tended to eliminate most of these features, increasing the vulnerability of local populations.

The diversity and complexity of the historical landscape also enabled greater flexibility in response to varying conditions, by allowing multiple life history options for native species. For example, because of the interplay of environmental factors, the most consistently productive salmon spawning and rearing habitats might be unfavorable in a given year; however, there would be other habitats that could be used.[4]

Native species today are particularly vulnerable to climate change because of the loss of these attributes. However, the potential exists to greatly increase the capacity of local ecosystems to accommodate climate change by recovering some of these characteristics of resilient landscapes. Enhancing groundwater recharge; enabling the natural capabilities of streams and wetlands to retain surface water; identifying, preserving, and expanding strongly perennial wetlands and stream reaches; recovering off-channel habitats such as side channels; reestablishing a valley oak canopy; and increasing the connectivity among habitats are all part of designing a more complex, robust, and resilient landscape. While the future is inherently unpredictable, local residents have the ability to choose alternative trajectories that will determine whether the Napa Valley's natural heritage will persist, thrive, and adapt in the future.

REDESIGNING THE VALLEY

People are always redesigning the valley. Today's Napa Valley is the product of the dreams, visions, and hard work of people 20, 50, 100, and even 500 years ago. Fifty years from now the valley again will look dramatically different, shaped by the visions of today. Examination of the past and present landscape shows that the valley's ecological health and resilience could be greatly enhanced within the framework of the contemporary developed landscape. It is not a choice between then and now, but a question of which elements of the valley people want to maintain and cultivate in the future.

What will the valley of 2050 look like? Will wood ducks and salmon occupy restored side channels of the river? Will we be graced by the beauty and shade of a new generation of valley oaks? What combination of floodplains and levees, wetlands and cities, willows and beavers will make the landscape the most pleasant to inhabit, the most economically viable, the most sustainable? With the perspective afforded by Kerr, Bartlett, Jepson, and other careful recorders of the past, we can begin to envision the landscape of the future.

Conceptual models of the Napa Valley's past, present, and potential future landscapes. (opposite page) A simplified schematic of the Napa Valley landscape shows the changes over the past two centuries, much of which had occurred by the 1940s. By 2012, some landscape features, such as tidal marshland, have been significantly revitalized. A vision for the future could include the reincorporation of valley oaks in patterns designed to provide ecological and societal benefits. River restoration, which has been relatively modest in comparison to historical river characteristics, could expand to include elements such as broad riparian forests, wetland complexes, and floodplain sloughs. Successful implementation of planned tidal marsh projects will continue to increase natural functions of the tidal lands. Bringing these elements together, local residents could develop a vision linking wetland, riverine, and terrestrial habitats to provide desired ecological and cultural functions.

8 • LANDSCAPE TOURS

Photographs by Ruth Askevold

Historical ecology helps us learn the forgotten stories of the land, the local narratives that reveal how nature and people have shaped our surroundings. These obscure details and subtle revelations are embedded everywhere we go, but they are generally hidden from our temporally limited view. When we can see the world within a historical landscape context, though, a boring drive is suddenly much more interesting. Street signs, dips in the road, and previously unnoticed trees are part of a living, dynamic landscape, a slow-motion drama outside our windows. Within our familiar landscape, we can recognize ecological loss and human ingenuity, threat and opportunity, alternative trajectories.

In this section, we present a selection of these landscape stories. The sites are not the most famous or historic spots in the Napa Valley (in fact, some are quite nondescript). But together these patterns contribute to a broader understanding of landscape patterns, the ecology they tend to support, and our cultural responses, so far, to this template.

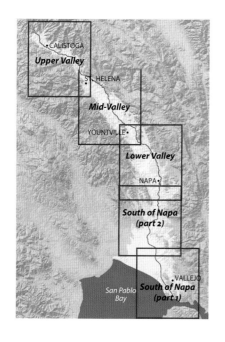

LANDSCAPE TOUR MAPS

The tour is divided into four parts, following the most common approach to the valley: from the south. Each of the maps—South of Napa (which is divided into two parts), the Lower Valley, the Mid-Valley, and the Upper Valley—is produced at the traditional scale of 1:62,500, in which 1 inch equals 1 mile. To juxtapose the reconstructed past with the navigable present, contemporary roads are overlaid upon historical habitats, with contemporary aerial photography as the background layer.

Studying the past from a moving car can be dangerous, so we recommend having a navigator and a driver, pulling over to observe from a safe place, walking, and other obvious precautions.

SOUTH OF NAPA: VALLEJO, AMERICAN CANYON, CARNEROS

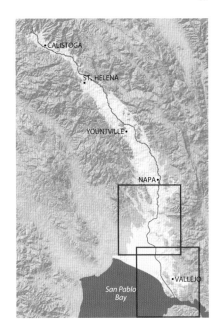

Crossing the Carquinez Bridge to approach the Napa Valley from the south, you travel over one of the great—albeit less appreciated—spans across San Francisco Bay. From this point, 148 feet above some of the deepest waters of the Bay, passengers can catch a first glimpse of the Napa River to the left, pinned close to land by two natural hills (one of them an island) and spanned by several bridges. This shoreline is the river's terminus and its confluence with San Pablo Bay, the broad embayment directly to the west. This is the entry point into the distinctive intertidal world and terrestrial margins south of Napa. Here an arc of incongruous cultural landscapes—Vallejo, American Canyon, unincorporated areas south of Napa, and the Carneros region—share the shores of the Napa baylands, their differences reflecting unique traits of the underlying landscape.

In its lowest 3 miles, the Napa River, after myriad other incarnations, flows through the constriction of Mare Island Strait, where it becomes a naval waterway that supported Navy shipyards from the 1850s until base closure in 1996. Here the proximity of Mare Island to the river's deepwater channel led to the establishment of the first U.S. Navy shipyard on the Pacific Coast and the development of the largest urban area alongside the Napa River—Vallejo. (With almost 116,000 people, Vallejo has a greater population than that of the entire rest of the watershed.) The shoreline history here, where tidal habitats have been filled to create an industrial shoreline, is more akin to that of San Francisco or Oakland than the rest of Napa Valley.

North of Mare Island Strait, the historical river was not naturally confined, and it extended more widely into a tidal landscape bordered primarily by seasonally flooded meadows. On the east side, from American Canyon to Napa, these heavy clay soils largely precluded the intensive agricultural development that happened to the north. Pastures predominated over orchards, leaving these lands open for more recent activities such as the Napa County Airport, industrial complexes, and suburbanization.

Across the river, the northwest margins of the baylands were better drained, permitting Pinot Noir and Chardonnay vineyards to thrive and take advantage of the cooling influence of the Bay waters. In these lands, as on the east side, relatively dense soils precluded most trees, except along creeks. South of Napa was a wide-open landscape of subtle topographic complexity, as well as abundant and diverse wetlands and uninterrupted views across meadows and marshes.

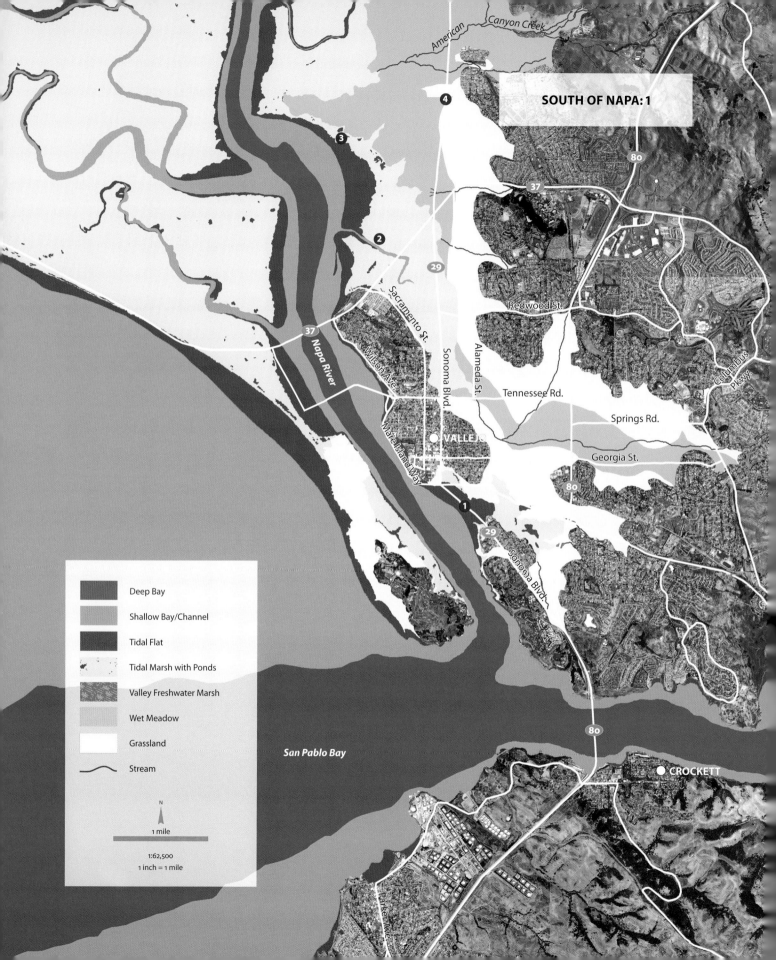

❶ VALLEJO SHORELINE:
A WORLD BENEATH THE PAVEMENT

To experience the extension of artificial land into the tidal Napa River, take the Sonoma Boulevard exit from Interstate 80. At Bennett Avenue, the road leaves the natural land surface and crosses onto Bay fill; the former tidal flats lie 10–20 feet below ground. At the Solano Avenue intersection, the historical shoreline was about 500 feet to the right and is still marked by a sudden climb in the road. Merging left onto Mare Island Way, the road briefly lands on the natural uplands on which old-town Vallejo was constructed, before Mare Island and the river become visible ahead. For the next mile (from just after Marin Street to a little past Tennessee Avenue) the road lies completely on Bay fill—a flat, novel landscape suspended above the former waters below. Across the river, piers and docks replace marshes. The rest of Mare Island Way to Highway 37 mostly occupies the foot of the natural hill called Vallejo Heights.

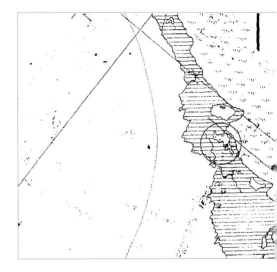

❷ WHITE SLOUGH:
RESTORED MARSHES ALONG THE RIVER

Heading north from the Napa River Bridge on Mare Island Way, Highway 37 passes an existing Napa River tidal marsh: the White Slough marshlands, which have been successfully restored since the late 1970s. While these brackish wetlands appear large from the road, they are a small portion of the historical extent, which continued for more than a mile and a half on the other side of the highway—and included the area with open water and buildings to the southeast.

❸ SLAUGHTERHOUSE POINT:
MARSHES AT THE EDGE

Along Highway 29 to the north, developed areas of Vallejo and American Canyon extend only partially into the historical tidal marshlands, providing an opportunity to see remnant and restored wetlands. A narrow strip of wetlands bordered Slaughterhouse Point to the south and west; these were all diked by the mid-20th century. Between 1985 and 1993, these marshes were restored as mitigation projects.[1] Despite several decades of separation from the tides, these suburban wetlands exhibit many features of mature tidal marshes, including two large marsh ponds along Meadows Drive near Severus Drive.

A tidal marsh pond mapped in 1856 can be seen today at Slaughterhouse Point.
Top: Rodgers 1856, courtesy of NOAA. Bottom: photograph April 1, 2010.

❹ BARTLETT'S WILDFLOWERS

Continuing north on the Napa-Vallejo Highway (Highway 29), the route retraces John Russell Bartlett's March 1852 trip. Here the itinerant bookseller and wayward Mexican Boundary Commissioner described "immense herds of cattle…feasting" on introduced wild oats while "wild flowers of varied hues were thickly scattered around." Bartlett had entered the extensive expanse of wet meadows and seasonal ponds south of Napa. The valley here was "perfectly level, without a hill or depression" and "still very wet." Looking closely at the contemporary land surface, one can see that recent buildings and infrastructure are slightly elevated on artificial fill above the surrounding lowland.

❺ GREEN ISLAND ROAD:
OLD SALT PONDS, NEW MARSHES, AND LANDLOCKED ISLANDS

In the absence of buildings and levees, Bartlett was able to see vessels "gliding up" the Napa River to the west, but from today's highway one has little sense of the river and baylands. Green Island Road follows a natural peninsula that provides access to the west (utilized also by the railroad). The road heads west from Highway 29 across former wet meadows; after about a mile, it jogs to the north to climb onto a series of low hills that extend into the marshlands. Green Island Road then continues west before turning sharply north. From this point on, the road follows the foot of the hills within a few hundred feet of the edge of the historical marsh. As the road approaches the next intersection, it bisects a finger of marshland that formerly extended inland along a low draw between hills. At the end of the road, one can see Green Island, the Napa River, and salt ponds restored to tidal action in 2009.

❻ IN THE WET MEADOWS

North of Green Island Road, Highway 29 briefly intersects a subtle intrusion of bedrock uplands that support the vineyards on the east side of the road. The broad, open wet meadows extended for a mile and a half to the west, where they merged into the tidal marsh ("In long, straggling, gleaming arms, the Bay died out among the grass"[2]). Some of the still-undeveloped areas around the airport give a glimpse of that landscape, described by Bartlett in 1852 ("it is an open plain, destitute of trees, and covered with luxuriant grass") and Stevenson in 1878 ("bald green pastures…the sun shone wide over open uplands"). A few freshwater sloughs occupied low swales meandering through the lowlands and draining to the marshlands. Showy Indian clover, a plant of wet swales, was collected here by Michener and Bioletti in 1891 and again by Callizo in 1952. Rare wetland species are still found in the area, such as tricolored blackbird, western pond turtle, and vernal pool fairy shrimp.[3]

Deep Bay

Shallow Bay/Channel

Tidal Flat

Tidal Marsh with Ponds

Valley Freshwater Marsh

Perennial Freshwater Pond

Wet Meadow

Vernal Pool Complex

Oak Savanna

Grassland

Stream

N
1 mile
1:62,500
1 inch = 1 mile

Soscol House in 1937 and 2006. *Top: courtesy of the California Historical Society. Bottom: photograph December 30, 2006.*

❼ SOSCOL FERRY ROAD:
FROM NATIVE LANDING TO HIGHWAY BRIDGE

The landscape shifts suddenly at this inconspicuous road. Not coincidentally, Soscol Ferry Road has a particularly dense and consequential history, including the brutal Battle of Suscol. This was the location of the Indian village and landing town of Soscol (variously spelled *Suscol*), located at the only significant constriction in the broad Napa River marshlands. Here a line of prominent bedrock hills rises from the soft lowlands, forming a natural causeway across the wet meadows and marshes, leaving a gap of just 1,000 feet in the center. Like Mare Island to the south, these hills provided easy, year-round access to the deepwater tidal channel, allowing landings even at low tide. Soscol was the main village for the Napa tribelet, providing access to fishing, waterfowl hunting, tule gathering, and other activities in the greater Napa marshlands.

In the late 1830s, 34 Indians were killed in a battle between Central Valley Indians and local Napa settlers (including perhaps some members of the Soscol tribe) and were buried in a tidal channel.[4]

In the 1850s, the Petaluma–San Francisco stagecoach had a ferry crossing here. At low tide, large steamships from San Francisco that were too big for the narrower river upstream stopped at Soscol, where passengers transferred to stagecoaches, smaller shuttle ferries, and later the railroad for the rest of the ride to Napa. Thompson's famous orchards, which demonstrated the success of dry farming the moist soils, were also located here. While there is markedly little evidence of all of this activity, the landforms remain, most recently occupied by the Butler Bridge, still known as the Southern Crossing. Old valley oaks, the edge of the historical marsh, and Suscol Creek can be seen from the road. And Soscol House, a roadside inn built by Elijah True in 1855 to serve travelers, persists in slightly modified form and location as the Italian restaurant Villa Romano near the freeway.[5]

❽ VISTA POINT:
200 YEARS x 360°

The top of the peak, publicly accessible from Soscol Ferry Road, is a rare high point near the baylands, providing a panorama across space and time. Two of the Bay Area's highest peaks, Mt. Tamalpais (directly southwest) and Mt. Diablo (nearly southeast), orient the view to the south. If you look towards Mt. Tam, you'll see a vantage point is similar to the 1869 Marple painting (see pages 120–121). From here, at different points in the 19th century, one would have seen tule canoes, flat-bottomed skows, and steamships navigating through the marshlands. Immediately north, the Napa Pipe industrial site hugs the river—marshes, oaks, and vernal pools were here a century ago. Across the river to the northwest, Bay waters reentered the south wetlands in 2001, creating a dynamic view that may be mostly open water or mudflat and marsh depending on the tide. Directly west, a line of eucalyptus trees marks the historical route to the other side of the ferry crossing.

9 STANLY LANE AND HORSESHOE BEND

After the bridge descends onto its bedrock footing, the road crisscrosses the historical edge of marsh until the stoplight at the Highway 12/29 juncture. Except for a few small areas adjacent to the road, all of the land inside the highway's curve is former tidal marshland surrounding the now-cutoff Horseshoe Bend (see page 132).

To the left, the line of eucalyptus trees follows higher ground along the old approach to the landing from the west. Stanly Lane followed the edge of this low rise more sinuously in the 1850s, was straightened by 1900, and is now a public path for hiking and biking (accessible from Highway 12 just west of the stoplight). The tall corridor of eucalyptus was a dramatic and persistent addition to the open, rolling hills and plains of the Carneros. In recent years, the trees have been thinned because of disease and replaced, in part, with valley oaks.

10 CUTTINGS WHARF LANDING

Once you've followed Cuttings Wharf Road through the agricultural Bay margins to the river's edge, the public landing provides a view of contemporary ship traffic on the river and the wetlands of Bull Island (see page 139) immediately across.

11 THE CARNEROS PLAINS

The Carneros area still maintains its differences from the rest of the Napa Valley, thanks to its distinctive soils and the cooling influence of the Bay. Oak savanna was naturally excluded by heavy subsoils and limited groundwater, and local trees continue to be rare; the few visible trees are relatively recent additions to the naturally open landscape. This unique area contributed certain ecological values. For example, the low herbaceous vegetation, seasonal ponding, proximity to the tidal marshlands, and absence of roosting sites for predatory raptors made the Carneros plains ideal rearing habitat for waterfowl. Stewart Duhig recollected: "My father told of running through the fields when he was a boy, about 1868–1875 in April and May and having to be careful not to step on any of the myriad of baby ducks and geese rising from their feeding grounds in such numbers they would darken the sun."[7] Some sense of this open landscape can still be gleaned along Duhig Road. Nearby, Carneros and Huichica creeks still maintain the riparian corridors that were particularly prominent against the surrounding meadows.

⓬ GIANT OAKS AT THE MARSH EDGE

North of the Vista Point hill, wet meadows were replaced by terrace lands—well-drained alluvium immediately adjacent to the tidal marshlands. Here valley oaks followed slightly higher ground right down to the marsh edge, creating an unusual transition directly between tidal wetlands and oak groves, in contrast to the open meadows that surrounded the marshlands elsewhere in Napa and most of the rest of San Francisco Bay. The commercial and industrial area of Rocktram and Napa Valley Corporate Park, west of the highway, was a complex ecotone of valley oak savanna, vernal pools, and tidal marsh. This is the largest area of tidal marsh fill in Napa County. A few large oaks are still visible.

At Kennedy Park, Napa Golf Course, and Napa Valley College, one can get a glimpse of the distinctive characteristics of this part of the east edge of the marshes. While the marshes are reclaimed, a number of oaks still mark the upland, and the natural drop-off to the tidal plain is evident. This would have been the downstream extent of riparian forest following the natural levees along the river.

Kennedy Park provides access to the Napa River Trail, which occupies the several-foot-high levee separating the park from the river. Following the trail north, one can see, on the west, the floodplains and terraces recently created to provide flood protection upstream and, on the east, the reclaimed baylands, which include all of the park and most of the adjacent Napa Golf Course. These areas are likely to be reclaimed by the tides, based on current sea level rise projections.[8] Native riparian vegetation has been planted along the levee as part of the Napa River Flood Protection Project.

⓭ NAPA GOLF COURSE:
HISTORICAL ECOLOGY FROM THE FAIRWAY

Built in 1968, Napa Golf Course was designed to take advantage of the historical landscape features that persisted at that time. Most of the course occupies former tidal marshland, which was diked by 1900 and converted to agricultural use in the 1960s. The course is known for its extensive water features—largely, a former tidal channel—and the prominent valley oak in the middle of the fairway on the 10th hole. Integration of historical landscape complexity gives the course its distinctive and challenging character.

⓮ IN AND OUT:
WHERE THE TIDES CAME

Tidal marshland extended farther into the valley than might be realized. The rising and falling tides continued through Kennedy Park and across Imola Avenue to within a few feet of In-N-Out Burger—an aptly named marker of the inland tidal extent.

Remnant valley oak at the Napa Golf Course. *Photograph September 11, 2009.*

LOWER VALLEY: NAPA, OAK KNOLL, YOUNTVILLE

As he approached the growing town of Napa in 1878, Robert Louis Stevenson noticed a shift in the landscape. The "hills began to draw nearer on either hand…soon we were away from all signs of the sea's neighborhood, mounting an inland, irrigated valley." As one leaves the Napa River's expansive delta and enters the more enclosed valley, terrestrial and riverine influences become more dominant. The river becomes narrower and more densely forested. It branches widely in parallel side channels created by flood events, rather than in tidal slough networks formed by daily tidal currents. Alluvial fans project from the western hills, pushing the river and its complex floodplain to the east side of the valley. An evenly spaced series of towns occupies the fans proceeding upvalley: Napa, Oak Knoll, and Yountville. While the open, foggy bayside landscape to the south supports Pinot Noir and Chardonnay, the well-drained, hot-summer savanna landscape more commonly helps to create Cabernets.

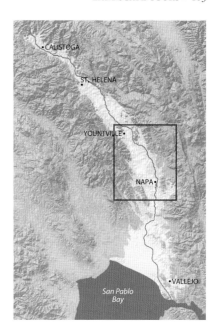

In the lower Napa Valley, early visitors first encountered grand oak savannas, lush year-round wetlands, broad riparian forests, and distinctive vernal pool complexes. Numerous creeks converged into the valley from the steep foothills. Many of them spread into seasonally wet meadows; others, such as Napa and Milliken creeks, had sufficient power to carve channels all the way to the river. Dry Creek, an important steelhead stream, created the Oak Knoll bench, one of the most important alluvial fans for grape growing.

While the Napa River channel is naturally much narrower here than it is in the tidal marshlands, the river still significantly defines the cultural landscape. The city of Napa closely hugs the river by historical necessity, for tidal access—with the helter-skelter alignment of bridges and street grids set by the river's sinuous course. North of the city, infrastructure is prudently located away from the river. Highway 29 and the Silverado Trail stick to the valley margins, following the traditional high roads above flooding; lateral crossroads are limited to a few places where the river is sufficiently well constrained for a viable bridge. The presence of only one crossing between Trancas Street and Yountville Cross Road (Oak Knoll Avenue)—a distance of 8 miles—is testament to the river's broad multi-channeled floodplain.

To approach the valley from the south today, flanked by the converging hills, is to revisit the grand entrance that has impressed many of the great chroniclers of the Napa Valley: "a fertile valley, enclosed by high mountain ridges, a rich bottom with grand trees, a stream rich in fish."[9]

⓯ JUAREZ ADOBE:
CHANGES AROUND NAPA'S OLDEST BUILDING

One of the most remarkable buildings in the Napa Valley is a modest restaurant at the junction of Soscol Avenue and the Silverado Trail. Built in 1840, Cayetano Juarez's adobe may have been the first in the Napa Valley and is now its oldest building. The adobe lies on the historic road between Vallejo and Napa, on the way to the early bridge across the river at Soscol Avenue. Not only has the durable building survived, but it has been, until recently, an active commercial establishment: the Old Adobe Bar and Grill (see following page).

⓰ THE VIEW FROM 3RD STREET BRIDGE

Within the Napa River channel, tidal influence continues upstream well beyond the last tidal marshes, giving rise to the landing town of Napa, located on the river's banks. From the 3rd Street Bridge, one can see the challenges of building so close to the river, and the recent accommodations that have been made to give the river more space and thereby reduce flooding. The 3rd Street and Soscol Avenue bridges have been rebuilt to provide more clearance. On the west side, a new trail runs between the new floodwalls and the tidal river, providing views that change by the hour, with mudflats exposed at low tide. On the east side, where a building on the northeast end of the bridge was removed, the river has formed a new gravel bar. Immediately upstream, at the end of 2nd Street, is where Frank Leach, remembering his childhood in the 1850s, described "the 'swimming hole' for the boys of pioneer days."[10]

Gravel bars were once common on lower Napa River. This bar formed soon after the river was widened to increase flood capacity and habitat. The gravel bar lies, in part, where a warehouse stood just a few years before. *Photograph by Susan Schwartzenberg, November 22, 2008.*

⓱ RECONNECTING A FLOODPLAIN SLOUGH

The 1st Street Bridge crosses both Napa Creek, just above its confluence with the Napa River, and a remnant segment of a former overflow channel of the river. Shown in Kerr's 1858 T-sheet, and still intact as a "Slough" in the late 19th century,[11] the channel carried high flows parallel to the river from a branch point over a mile upstream. The remnant portion (which currently exhibits an odd array of plants, including palm trees) will be expanded and reconnected to the river as part of the Oxbow flood bypass, reestablishing some aspects of its historical function.

⓲ FISHING THROUGH A HOLE IN THE FLOOR

In 1910, the Main Street business district continued straight across Napa Creek. Buildings suspended over the creek completely covered the channel between Pearl and Main streets—a length of 700 feet.[12] There used to be a sporting goods store over the creek; longtime residents remember fishing through the trap door to the creek.[13] As part of redevelopment in the 1980s, the river was uncovered.

The Juarez adobe. Shown in 1895 and 2006. *Top: courtesy of The Bancroft Library, UC Berkeley. Bottom: photograph December 26, 2006.*

⑲ THE OXBOW

The unusual sinuousity of "the Oxbow" limited early development of the river's margin, enabling several public access points today. The serpentine river course can be seen from the Oxbow Public Market and adjacent shoreline trail. On the east side of the river, the natural peninsula formed by the upstream meander is now the Oxbow Preserve City Park (accessed from MacKenzie Drive off of Silverado Trail). The park displays a complex floodplain topography comprising benches, old channels, seasonal ponds, and older surfaces occupied by older oaks. Yet 70 years ago there were no trees on most of this land, except a few on the very edges of the banks; the entire forest has regenerated in recent decades. The prow of the park is lined largely by walnut trees, probably washed downstream from the orchard-era valley.

⓴ TRANCAS CROSSING PARK

Trancas Crossing Park, opened in 2011, provides direct access to the Napa River as part of a natural floodplain park. Because of its position at the upper limit of tidal waters, the site has been the location for Native American settlements, an important Mexican-era landing, a diversion dam providing water for the city of Napa, and a major river crossing. While the park was farmed for some decades, it was never leveed; floods can briefly submerge the area beneath 10 or more feet of flowing water. On the east side of the park, the Napa River Dam impounded the river for nearly a half century (from the 1880s to around 1930[13]), potentially playing a significant role in the decline of Napa's fisheries. As part of the new park, artists Elise Brewster and Josanna Borelli-Zavala created a 12-foot-high flood marker demonstrating the depth of inundation during recent and historical floods. At the park, interpretive signage based on this atlas explores these topics further.

㉑ FROM FLOODPLAIN TO FAN:
TRANCAS STREET BRIDGE

The bridge and park are good places to witness the subtle topography of an alluvial fan. The steep banks on the west side of the park are the result of the Napa River cutting into the alluvial fan of the Dry Creek/Napa Creek drainages. In contrast, the east side of the park joins the lower land of the Napa River floodplain. This elevational difference can be seen in the upward slope of the Trancas Street Bridge as it heads west from floodplain to fan.

㉒ SIDE CHANNELS ALONG SILVERADO TRAIL

The riparian corridor on the east side of Silverado Trail north of Trancas Avenue is a former overflow channel of the Napa River, now disconnected and only receiving flow from small tributaries to the east. Just north of Hardman Avenue, a small bridge in the road (note the guardrail) marks the channel's former route from the river.

㉓ THE GIGANTIC OAKS OF THE DRY CREEK FAN

Highway 29 imperceptibly ascends the massive Dry Creek fan on the outskirts of Napa, after Salvador Avenue. The town located at the road's apex—Oak Knoll—is where Bartlett first observed that the valley was "now studded with gigantic oaks." Some of these trees can still be seen in narrow, untended spaces along the road, giving most travelers their first glimpse of the valley's great oaks, just as they did for Bartlett. Remnant trees can be seen from the frontage road, Solano Avenue, which is accessible from Oak Knoll Avenue, and at nearby Trefethen Family Vineyards.

㉔ THE WILLOW PATCH

North of Ragatz Lane, Highway 29 descends into an "interfluve," a low area between adjacent alluvial fans. Here there were hundreds of acres of seasonal wetlands and a year-round tule marsh an estimated 80 acres in size. In the late 1800s, the area was known as "the willow patch" and famous for amphibians: "During summer evenings the combined croaking of this community of frogs could be heard throughout the town."[14] Remnants of the marsh can still be seen in the Yountville Golf Course and adjacent winter-flooded fields. To the east was the extensive wetland complex along the Napa River at the mouth of Dry Creek, which included "The Willows"—likely the village site where Indians who worked for George Yount lived ("about a hundred…made their home near him in a willow copse about a mile distant"[15]).

㉕ PIONEER CEMETERY AND ANCIENT INDIAN BURIAL GROUNDS

Few sites in the valley explicitly acknowledge the underlying indigenous history. At the George Yount Pioneer Cemetery and Ancient Indian Burial Grounds, 19th-century settlers (including Yount) are buried along with generations of the Caymus tribe. The deep history of this small hill is reflected in some remarkable manzanita and live oaks, which, given their age and location, may have been planted by the Caymus.

㉖ NAPA RIVER ECOLOGICAL RESERVE

This State Department of Fish and Game reserve is a rare public place to experience the Napa River today. The site includes both the river and a former side channel, which converge to form a forest of valley oaks, live oaks, bay laurel, and other trees. The side channel is now disconnected from the river at its upstream branch point and serves only as the downstream extension of the tributary Conn Creek. Even at this preserve, the west side of the river is constrained by a major levee, originally constructed in the 1870s[16] and now colonized with mature oak trees.

Napa River Ecological Reserve. (above)
Photograph April 24, 2011.

The yellow-breasted chat (opposite page) was relatively common at the Napa River Ecological Reserve as recently as the 1980s.[17] *Painting by Andrew Jackson Grayson, courtesy of The Bancroft Library, UC Berkeley.*

MID-VALLEY: OAKVILLE, RUTHERFORD, ST. HELENA

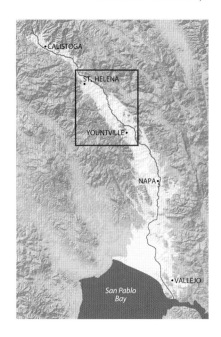

The Napa Valley constricts slightly at Yountville, as the Yountville Hills—a series of bedrock knolls—reduce the valley's effective width and divert the Napa River from side to side. Farther north, in the Rutherford and Oakville districts, the valley widens again into a diverse landscape that might be considered the heart of the Napa Valley. In this area there were both prominent alluvial benches and a broad floodplain, enabling the full expression of alluvial topographic complexity in ecological patterns. Landscape features included broad riparian forest on unconstrained side channels, natural depressions between converging fans, and slight rises that made ideal oak lands. As a result, the mid-valley had the greatest proportion of oak savannas, riparian forests, and valley wetlands. Not coincidentally, it has also been the source of many of the Napa Valley's most iconic and influential vineyards, from To Kalon to Niebaum-Coppola.

This central portion of the valley also receives two of the most important tributaries to the Napa River, Conn and Sulphur creeks. By geologic happenstance, these tributaries drain, from opposite sides of the valley, the two largest areas of erosive Franciscan geology in the watershed, contributing a great proportion of the coarse gravels needed to make downstream spawning beds for salmon.

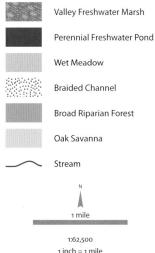

While Conn Creek is now dammed, the braided Sulphur Creek channel is still fairly natural. Its broad alluvial fan, the valley's largest, also signals another landscape shift. For the 4 miles north of Bale Slough, Sulphur Creek's fan exerts an overwhelming influence on the valley, covering almost its full width and forcing the Napa River against the east-side hills. Floodplain width is here reduced, giving way to higher ground supporting oak savannas, the town of St. Helena, and soils ideal for orchards and vineyards.

㉗ YOUNT MILL ROAD AND YOUNTVILLE HILLS

Yount Mill Road follows the base of the Yountville Hills between the town of Yountville and Highway 29, providing an up-close view of one of the Napa Valley's prominent knolls. Some of these isolated bedrock hills are surprisingly large, standing out "like green islands in an ocean of level verdure."[18] (Silverado Trail also passes through a number of knolls south of Yountville Cross Road.) A historical marker for Yount's flour mill is located just across the river on Cook Road. According to Yount's journals, the Caymus tribe initially resisted helping him construct the dam in the 1840s because the area was an important spiritual site, the dwelling place of the protector spirit of the Caymus tribe.[19] An accord was apparently reached that the Spirit would move across the creek to the adjacent hillside. The mill was powered by the Napa River through a dam and mill race,[20] suggesting a much shallower channel than that observed today.

㉘ LIVE OAK GROVE ON THE VALLEY FLOOR

The regional headquarters for the State Department of Fish and Game is located in an unusual stand of dense live oak woodland. The trees, also visible in early aerial photography, appear to be associated with the alluvial deposits of nearby Rector Creek. While valley oaks are the more common tree on the valley floor, this live oak stand may hint at some of the additional, undocumented complexity of historical valley vegetation.

㉙ CROSSING THE FLOODPLAIN

South of Rutherford Road, Conn Creek ceased to be a well defined channel, spreading into seasonal wetlands and joining overflow channels of the Napa

Near Money Road, valley oaks tower over vineyards and mark the route of a former side channel. *Photograph December 30, 2006.*

River in a broad combined floodplain. Oakville Cross Road spans this area, revealing some remnants of the recent floodplain, including the straight channel constructed to carry Conn Creek flows and evidence of several former side channels, including along Money Road.

③⓪ FROM GRAIN TO GRAPES

While grapes currently dominate the valley, a number of crops have been grown widely over the past 150 years. For example, in the 1870s this area was mostly grain fields. Thompson's farm (now largely the Niebaum-Coppola estate) was described as "largely devoted to wheat, of which he produces large crops, the soil being of the rich bottom land. There are also several acres of choice vines."[21]

③① BALE SLOUGH WETLANDS

The towns of Rutherford and St. Helena sit on adjacent alluvial fans. The low area where they converge was one of the largest freshwater wetlands in the valley, reflected in the name Bale Slough. While the area has been intensively ditched and channelized, it continues to form seasonal wetlands. The area where Highway 29 crosses the slough (now a larger creek channel) is thought to have been the first concrete roadway in the valley, necessitated because of the persistent flooding.[22]

③② ZINFANDEL OAK TUNNEL

Ancient valley oaks arching over the roadway near Zinfandel Lane give Highway 29 one of its most distinctive segments. The trees occupy the narrow, undeveloped margin between paved surface and vineyard.

Remnant wetlands of the former Bale Slough wetland complex. *Photograph June 15, 2007.*

33 ZINFANDEL LANE STEELHEAD POOL

One of the best places to see returning steelhead as they attempt to ascend the Napa River to upstream spawning grounds is the deep pool immediately downstream of the Zinfandel Bridge. A concrete apron underneath the historic stone bridge has protected the footings from erosion. The river downstream has continued to incise, however, creating an abrupt shift in elevation at the bridge, which is being modified in 2011 to improve fish passage. During the winter run, steelhead rest here and wait for suitable flow conditions to make the difficult jump over this barrier.

34 SULPHUR CREEK:
A DYNAMIC BRAIDED CHANNEL

The unique geomorphology and vegetation patterns of Sulphur Creek—an active, undammed, and now-unmined braided channel—can be seen from the South Crane Avenue/Valley View Street bridge crossing. With the cessation of in-channel gravel mining in 1999, and the addition of a natural sediment supply from the Franciscan headwaters, Sulphur Creek is a rare aggrading channel in a network of largely incising systems.

35 TAYLOR'S REFRESHER:
REROUTING OF SULPHUR CREEK

While much of Sulphur Creek's braided channel reach remains intact, the reach immediately above Main Street has been dramatically modified. In the early 1940s, the stream flowed in a large meander that paralleled Main Street. By 1949, when Taylor's Refresher (now Gott's Roadside Tray Gourmet) was established, this reach had been diverted into a narrow, straight channel.

Mature oaks amid younger tree plantings at the Charles Krug winery. *Smith and Elliot 1878.*

36 ST. HELENA

The early town took advantage of the shade of valley oaks, a few of which can be found scattered among yards along the roads. Many more could be strategically reintroduced to the townsite that Smith and Elliot described in 1878 as "[a] natural location, surrounded with an abundance of native trees." The village of the Indian Chief Canijolmanok was located in the vicinity of St. Helena, according to Salvador Vallejo's 1861 land-case testimony.

37 AN OLD OAK GROVE

Charles Krug, like many of the early wineries, was established in a natural oak grove. The stand remains substantially intact, now integrated with a complex of buildings, roads, and newer plantings.

UPPER VALLEY: LARKMEAD, CALISTOGA

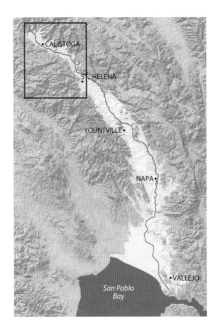

North of St. Helena the valley begins to narrow and, above Bale Lane, rotate to the west. The river continues to be pushed from one side of the valley to the other by the opposing alluvial fans of its tributary creeks. But instead of extensive floodplain wetlands and widely branching side channels, the river corridor is more restricted. We see evidence of a less broad riparian forest hugging the channel. In this narrower part of the valley, wet meadows were limited in the historical landscape. Yet perennial wetlands were found in places with distinct hydrology, such as the natural dam made of neighboring knolls at Dunaweal Lane, and the Calistoga hot springs. Textual accounts suggest that in this upper part of the valley, conifers such as Douglas fir and ponderosa pine joined the oaks as valley floor trees.

Two of the most well-documented perennial creeks, Bale and Ritchey, flow into the valley from the southwest, but these creeks do not appear to have flowed directly into the Napa River. The natural levees of the river blocked most of the creeks between St. Helena and Calistoga from direct confluence, so the creeks probably only connected during high-flow events, when steelhead, water, and sediment would make the leap between creek and river.

Two of the distinguishing characteristics of this part of the valley directed its 19th-century development. The thermal mineral waters led Calistoga to become a major resort destination by the 1860s, which it remains today. The reliable flows on Bale Creek supported one of the earliest important mills, drawing commerce from throughout the region. The mill, famous in the first half of the 20th century as an abandoned point of interest, is now a functioning water-powered grist mill, and the area is part of Bothe–Napa Valley State Park.

38 BALE MILL, MILL CREEK, AND LYMAN'S GROVE

Several distinct landscape characteristics can be experienced in robust form at Bale Grist Mill State Historic Park and the much larger, adjacent Bothe–Napa Valley State Park. Mill Creek and much of its watershed are accessible through public trails, as is the mill, built of local timbers from the surrounding forests in 1846 and originally powered by the stream's energy. Across the road from the mill lies the unique 1-square-acre Lyman Memorial Buckeye Grove, donated by William Lyman as a nature preserve.

39 OLD FAITHFUL GEYSER: LANDSCAPE CHANGE IN THE BACKGROUND

As a well-documented site, the geyser at Calistoga shows that certain landscape features can be remarkably stable; and some major changes, subtle. The still-reliable geothermal eruption in the foreground and the distinctive vegetation patterns in the hillside background overshadow dramatic changes on the valley floor. The view captured in a postcard circa 1920—across an open meadow, with several valley oaks in the distance—has since been filled with imported vegetation, obscuring the bottom half of the mountain. Yet it appears possible to identify individual trees and bushes on the visible slopes, their location unchanged by the passing of a century. Upon closer examination, however, the forest silhouette on the right flank of the mountain has apparently disappeared, probably the result of logging, and not returned.

Buckeye trees intermix with valley and live oaks at the Lyman Memorial Buckeye Grove, a rare undeveloped portion of the valley floor located alongside Mill Creek at the foothill margin. In early summer, the profusion of sweet-scented buckeye flowers suffuses the site with fragrance. *Photograph June 20, 2010.*

The geyser. *Left: postcard circa 1920. Right: photograph: June 16, 2007.*

④⓪ LOCAL WETLANDS

Downstream of the geyser, the spring water flows into a relatively intact marsh. The wetland supports a number of rare plant species and gives a glimpse of the distinct color, texture, and sounds of a natural marsh. The lawns of some adjacent properties are comprised largely of saltgrass, reflecting the high salt content of the spring-influenced soils.

④① NAPA RIVER AT THE SHARPSTEEN

The densely wooded river can be forded via a staircase and concrete walkway adjacent to the Sharpsteen Museum, which exhibits historical photographs of Calistoga in one of the original hot springs resort cottages. Oak trees extend into Pioneer Park on the opposite bank. The depth of the river channel is impressive considering Adams' description of early settlers needing to excavate a true channel for the multi-thread river system—creating "quite a creek running through our little city."[23]

④② OAKS GONE AND RETURNED

Photographs by Watkins and Muybridge memorialize the extensive stand of oaks in Calistoga that was almost completely removed by 1939. Within the city limits both old giants and younger descendants can be found.

Streetside valley and live oaks in downtown Calistoga. *Photograph June 16, 2007.*

COMMON AND SCIENTIFIC NAMES OF SPECIES

PLANTS

Alkali milkvetch (*Astragalus tener* var. *tener*)

Arroyo willow (*Salix lasiolepis*)

Basket sedge (*Carex barbarae*)

Big-leaf maple (*Acer macrophyllum*)

Bracted popcorn flower (*Plagiobothrys bracteatus*)

California bay laurel (*Umbellularia californica*)

California black oak (*Quercus kelloggii*)

California blackberry (*Rubus ursinus*)

California boxelder (*Acer negundo*)

California brickell bush (*Brickellia californica*)

California buckeye (*Aesculus californica*)

California lilac (*Ceanothus* spp.)

California poppy (*Eschscholzia californica*)

California rose (*Rosa californica*)

California sycamore (*Platanus racemosa*)

California wild grape (*Vitis californica*)

Calistoga popcorn flower (*Plagiobothrys strictus*)

Coast live oak (*Quercus agrifolia*)

Delta tule pea (*Lathyrus jepsonii* var. *jepsonii*)

Douglas fir (*Pseudotsuga menziesii*)

Douglas' meadowfoam (*Limnanthes douglasii*)

Dwarf calicoflower (*Downingia pusilla*)

Fremont cottonwood (*Populus fremontii*)

Goldfields (*Lasthenia* spp.)

Gray pine (*Pinus sabiniana*)

Hardstem tule (*Schoenoplectus acutus*, syn. *Scirpus acutus*)

Hazelnut (*Corylus cornuta* var. *californica*)

Large-flowered star-tulip (*Calochortus uniflorus*)

Lupine (*Lupinus* spp.)

Madrone (*Arbutus menziesii*)

Manzanita (*Arctostaphylos* spp.)

Maroonspot calicoflower (*Downingia concolor*)

Mason's lilaeopsis (*Lilaeopsis masonii*)

Narrowleaf willow (*Salix exigua*)

Nutsedge (*Cyperus eragrostis*)

Oregon ash (*Fraxinus latifolia*)

Pacific gumplant (*Grindelia stricta*)

Parry's tarplant (*Centromadia parryi* ssp. *parryi*)

Pickleweed (*Salicornia virginica*, syn. *Sarcocornia pacifica*)

Ponderosa pine (*Pinus ponderosa*)

Red alder (*Alnus rubra*)

Red willow (*Salix laevigata*)

Rusty-haired popcorn flower (*Plagiobothrys nothofulvus*)

Sack clover (*Trifolium depauperatum*)

Saltgrass (*Distichlis spicata*)

San Joaquin spearscale (*Atriplex joaquiniana*)

Shining willow (*Salix lucida*)

Showy Indian clover (*Trifolium amoenum*)

Slender popcorn flower (*Plagiobothrys tenellus*)

Smartweed (*Polygonum punctatum*)

Soft bird's beak (*Chloropyron molle* ssp. *molle*; syn. *Cordylanthus mollis* ssp. *mollis*)

Tidy-tips (*Layia platyglossa*)

Toothed calicoflower (*Downingia cuspidata*)

Valley oak (*Quercus lobata*)

Yellow monkeyflower (*Mimulus guttatus*)

Yellow willow (*Salix lutea*)

ANIMALS

Acorn woodpecker *(Melanerpes formicivorus)*

Arrow goby *(Clevelandia ios)*

Bay goby *(Lepidogobius lepidus)*

Beaver *(Castor canadensis)*

Black-tailed deer *(Odocoileus hemionus)*

California clapper rail *(Rallus longirostris obsoletus)*

California roach *(Lavinia symmetricus)*

Chinook salmon *(Oncorhynchus tshawytscha)*

Chum salmon *(Oncorhynchus keta)*

Cinnamon teal *(Anas cyanoptera)*

Coastrange sculpin *(Cottus aleuticus)*

Coho salmon *(Oncorhynchus kisutch)*

Delta smelt *(Hypomesus transpacificus)*

Grizzly bear *(Ursus arctos horribilis)*

Hardhead *(Mylopharodon conocephalus)*

Hitch *(Lavinia exilicauda)*

Jack smelt *(Atherinopsis californiensis)*

Longfin smelt *(Spirinchus thaleichthys)*

Longjaw mudsucker *(Gillichthys mirabilis)*

Mallard *(Anas platyrhynchos)*

Northern anchovie *(Engraulis mordax)*

Pacific herring *(Clupea pallasii)*

Pacific lamprey *(Lampetra tridentata)*

Pacific pallid bat *(Antrozous pallidus pacificus)*

Pacific staghorn sculpin *(Leptocottus armatus)*

Prickly sculpin *(Cottus asper)*

Pronghorn *(Antilocapra americana)*

Rainbow trout *(Oncorhynchus mykiss)*

Riffle sculpin *(Cottus gulosus)*

River lamprey *(Lampetra ayresi)*

Ruddy duck *(Oxyura jamaicensis)*

Sacramento blackfish *(Orthodon microlepidotus)*

Sacramento perch *(Archoplites interruptus)*

Sacramento pikeminnow *(Ptychocheilus grandis)*

Sacramento splittail *(Pogonichthys macrolepidotus)*

Sacramento sucker *(Catostomus occidentalis)*

Salt marsh harvest mouse *(Reithrodontomys raviventris)*

Shiner perch *(Cymatogaster aggregata)*

Sockeye salmon *(Oncorhynchus nerka kennerlyi)*

Speckled sanddab *(Citharichthys stigmaeus)*

Starry flounder *(Platichthys stellatus)*

Steelhead *(Oncorhynchus mykiss)*

Striped skunk *(Mephitis mephitis)*

Thicktail chub *(Gila crassicauda)*

Three-spine stickleback *(Gasterosteus aculeatus)*

Tidewater goby *(Eucyclogobius newberryi)*

Tricolored blackbird *(Agelaius tricolor)*

Tule elk *(Cervus canadensis ssp. nannodes)*

Tule perch *(Hysterocarpus traski)*

Vernal pool fairy shrimp *(Branchinecta lynchi)*

Western brook lamprey *(Lampetra richardsoni)*

Western pond turtle *(Actinemys marmorata)*

White sturgeon *(Acipenser transmontanus)*

White-breasted nuthatch *(Sitta carolinensis)*

Wood duck *(Aix sponsa)*

Yellow-billed cuckoo *(Coccyzus americanus)*

Yellow-billed magpie *(Pica nuttalli)*

Yellow-breasted chat *(Icteria virens)*

NOTES

1 • EXPLORING THE NAPA VALLEY

[1] Hamilton (1997), Foster (2000), and Montgomery (2008). Foster (2000) noted, "It is often overlooked that many fundamental insights into ecological processes and major environmental issues come not through reductionist or high tech studies of modern conditions but from thoughtful consideration of nature's history."

[2] Rackham (1994) warns of "pseudo-history"; Foster and Motzkin (2003) show the potential for misinterpretation of historical landscapes; and Harley (1990) reminds us that all historical documents are incomplete and biased.

[3] Overviews of historical ecology theory and techniques are provided by Swetnam et al. (1999), Egan and Howell (2005), and Balee (2006).

[4] Sauer (1930).

[5] Sauer (1939). This introductory section to the original version of *Man in Nature* was titled "The Chapter You Should Write: These are some of the things you will want to learn about studying your part of the country as it was in Indian days."

[6] See discussions by Foster (2002), Stewart et al. (2002), and Montgomery (2008).

[7] For environmental histories, see Cronon (1983), Worster (1985), and White (1995). Analyses of riverine systems include Petts et al. (1989), Collins et al. (2003), Kondolf et al. (2007), and Walter and Merritts (2008). Ecological reconstructions of terrestrial vegetation change include Bahre (1991), Rackham (1994), Whitney (1994), Manies and Mladenoff (2000), Hulse et al. (2002), Bolliger et al. (2004), and Sanderson (2009).

[8] Goals Project (1999).

[9] See Sloan (2006) for a discussion of Bay Area geology and Swinchatt and Howell (2004) for the geologic origin of Napa Valley.

[10] As recently as 1960, a water resources study noted, "Because the rainfall is relatively high in winter and because nights in summer are cool and moist, prunes and grapes mature without irrigation. Accordingly, no vineyards are irrigated, and only about 1,500 acres of orchard and about 2,000 acres of permanent pasture are irrigated [in Napa and Sonoma valleys]" (Kunkel and Upson 1960).

[11] Lebassi et al. (2009) and Cayan et al. (2011).

[12] The ranges adjacent to the Napa Valley are sometimes considered part of the Mayacamas Mountains (on the west) and the Vaca Mountains (on the east), but we use the more locally specific Mt. Hood Range and Napa Range here.

[13] Sloan (2006).

[14] Swinchatt and Howell (2004).

[15] Ibid.

[16] Carpenter and Cosby (1938) and Helley et al. (1979).

[17] Swinchatt and Howell (2004).

[18] Milliken (1995, 1978), Cook (1956), and Dillon (2001). Milliken (1978) notes that the disparate available sources leave "a bewildering assortment of names for rancherias, groups and dialects"; these names and locations follow Milliken's integration. For a lengthier discussion of California prehistory, see Moss and Erlandson (1995), Erlandson and Jones (2002), and Lightfoot and Parrish (2009).

[19] Milliken (1978).

20 Stewart et al. (2002), Anderson (2005), and Lightfoot et al. (2009).

21 For the recovery of traditional ecological knowledge, see Anderson (2005) and Cuthrell et al. (2009).

22 Milliken (1995: 10).

23 Milliken (1978) and Milliken (1995: 237–8, 248).

24 Sources for timelines include Menefee (1873), Husmann (1912), Carpenter and Cosby (1938), Milliken (1978), Weber (1997, 2001), Sullivan (2008), the State Census Data Center (2010), and Noonan (2011). Major land-use phases shown are conceptually based on a variety of qualitative and quantitative data and should not be considered precise.

25 For example, Father Altimira observed burned land in the Napa River watershed in 1823 (Altimira 1861 and Smilie 1975), and Vallejo signed a treaty with the Wappo to restrict their burning in 1836 (Smilie 1975).

26 Bartlett (1854: 18).

27 Olmstead and Rhode (2003) and Johnston and McCalla (2004).

28 Husmann (1912) and Carpenter and Cosby (1938: 8). Carpenter and Cosby noted that the steepest decline in grain cultivation occurred during 1880–1910.

29 Sullivan (2008).

30 Lapham (1949).

31 Poland and Ireland (1988).

32 The total includes Calistoga, St. Helena, Yountville, Napa, and American Canyon. The city of Vallejo (115,942 in the 2010 census) also lies at the very southern edge of the watershed.

33 Altimira (1861).

34 *San Francisco Call and Post* (1921) and Fels (2005).

35 Jepson (1934: 68).

36 Grossinger and Beller (2011); Beller et al. (2010).

37 Santa Clara Valley imagery from the Coyote Creek Watershed Historical Ecology Study (Grossinger et al. 2006).

38 SFEI (2011a).

39 A Living River approach seeks to balance a broad array of ecological and societal needs such as flood protection, dynamic natural function, and ecosystem protection (van der Velde et al. 2006).

40 For more discussion of synthesis methods, see Grossinger (2005), Grossinger et al. (2007), and Whipple et al. (2011).

41 Despite decadal-scale variability, climatic characteristics during the period from which historical data were obtained (1820s to 1940s) were relatively stable (Dettinger et al. 1998).

42 Grossinger (2005) and Grossinger et al. (2007).

43 Malamud-Roam et al. (2006, 2007).

44 Goals Project (1999).

45 Goals Project (1999: "Moist Grassland") and Grossinger et al. (2007).

46 Holland (1986), Sawyer et al. (2009: "Salt Grass Flats"), and Grossinger et al. (2007).

47 Cowardin et al. (1979), Holland (1986), and Mayer and Laudenslayer (1988: "Fresh Emergent Wetland").

48 Cowardin et al. (1979).

49 Collins et al. (2006).

⁵⁰ Graf (2000) and Beechie et al. (2006).

⁵¹ Minnich (2008) and Sawyer et al. (2009: 30–31).

⁵² Allen et al. (1991: "Valley Oak–Coast Live Oak/Grass") and Sawyer et al. (2009: "Valley Oak Woodland").

2 • OAK SAVANNAS AND WILDFLOWER FIELDS

¹ Bartlett (1854).

² *Daily Alta California* (1860). The article also noted that "the trees in the valley are not very numerous, and were long used under the Mexican dominion as shelters from the sun by the wild cattle, and thus the soil beneath them was enriched by manure." See also Jepson (1909) and Bartlett (1928).

³ Smith and Elliott (1878: 3): "All kinds of oaks that grow are brittle and worthless…almost all supplies of oak, ash and hickory are brought from the East." Peattie (1953: 422–423): "In contrast to the grand stature of the Valley Oak is the uselessness of its wood." Longtime resident Babe Learned described the gradual cutting of valley oaks for firewood in St. Helena (interview by Elise Brewster, October 5, 2002).

⁴ Whipple et al. (2011) and Shortridge (1896: 184, 192).

⁵ The steep decline in wheat cultivation at the end of the 19th century and the parallel expansion of orchards serve as proxies for the conversion of oak savanna pastures and fields to intensive agriculture.

⁶ California State Senate (1859).

⁷ Evans (1873).

⁸ *Daily Alta California* (1857).

⁹ Valley oaks were extensively described in the Santa Clara Valley, which was called "Plain of the White Oaks" *(Llano de los Robles;* see Beller et al. 2010). Extensive wet meadows and chaparral stands on Pleistocene alluvial fans limited the extent of oak savanna.

¹⁰ We examined the correlation between oak distribution and soil types to identify the most suitable soils for oaks, and we used those soil types to help define savanna extent (e.g., Whipple et al. 2011). This analysis indicated that oaks were consistently associated with loam soils, particularly gravelly loams, and rare in clay soils. We expect that the mapped savanna also includes some areas with few or no trees.

¹¹ Carpenter and Cosby (1938: 40) and Lambert and Kashiwagi (1978: 23–24).

¹² Contemporary, non-riparian valley oaks with trunk circumference at breast height greater than 3.0 meters (0.94 meters/ 3 feet calculated diameter) were mapped using GPS from February to August 2002. (The census was led by Jake Ruygt and Stephen Rae, and conducted through the Friends of the Napa River.) Most of these valley oaks were likely present before significant landscape modification. More than half the trees had diameters greater than 4 feet; almost 100 were larger than 5 feet in diameter. Valley oak age versus trunk size relationships can vary substantially based on growing conditions, but average estimates are in the range of four to six years per inch of diameter (based on ten cores of Napa Valley trees carried out during this project); stump measurements from Sonoma Valley (Arthur Dawson, pers. comm.); Jepson (1910); see also Pavlik et al. (1991: 10). The smallest of these trees would thus be expected to be, on average, 148 to 222 years old (germination during 1780–1854). At four years per inch, about half of the data set would have been present 200 years ago; at six years per inch, all of the trees would have been.

¹³ Some large valley oaks remained in fields, rangelands, and, to a lesser degree, orchards in the 1940s, providing evidence of earlier distribution. Trees not already mapped by GPS were digitized based on their distinctive size, shape, and groupings (Brown and Davis 1991, Brown 2002, Sork et al. 2002, and Whipple et al. 2011).

[14] Kerr (1858), Dewoody (1866a), Mow (1868), Morgan (1871), and Dewoody (1873). Near Calistoga, three partially overlapping maps showed a total of 500 oaks in their area of overlap, some of which were likely the same tree depicted twice. No other regions were mapped by overlapping sources, so even allowing for maximum redundancy, these maps provide evidence for over 900 individual trees.

[15] Sources include lithographs from Smith and Elliott (1878); landscape photography by Carleton Watkins, Eadweard Muybridge, W. W. Lyman, and Turrill and Miller; and descriptions by Bartlett (1854), California Senate (1859), Taylor (1862), Smith and Elliott (1878), and Jepson (1934: 67).

[16] Scattered: Thompson (1857), Brewer (1861, pub. date 1966), Evans (1873), Menefee (1873), and Smith and Elliott (1878). Scattering: Gray (1853). Dotted: *Daily Alta California* (1857), McClellan (1872), Evans (1873), Menefee (1873), and Smith and Elliott (1878).

[17] Definitions of savanna vary but generally involve canopy cover of less than 25% and a low herbaceous understory (FGDC 1997 and Allen-Diaz et al. 1999). Whipple et al. (2011) found in Santa Clara County that the GLO terms "scattered" or "scattering" corresponded to savanna densities. While sufficient GLO data to calculate stand densities are not available in the Napa Valley, some evidence can be gleaned from historical maps and photographs. Oak lands shown by Kerr (1858), prior to intensive agriculture, or visible in pasture lands in 1942 aerial photography, achieved densities of 0.3 to 0.8 trees per acre across 250- to 600-acre patches. (Even here, the trees were not evenly spaced.) This range of densities is similar to stand densities calculated for valley oak savannas in the Santa Clara Valley (0.7 to 1.5; Whipple et al. 2011) and Santa Ynez Valley (0.6; Sork et al. 2002), but substantially less than that reported for denser woodland areas of the Santa Clara Valley (Whipple et al. 2011). (The denser Santa Clara County areas were also described with descriptors not found in the Napa Valley, such as "thick woods" or "woodland.")

[18] Bartlett (1854).

[19] Smith and Elliott (1878) and Stevenson (1884).

[20] Bowles (1865) and Cronise (1868: 180).

[21] Turrill (1876), Brewer (1861, pub. date 1966), Brace (1869), Avery (1878), Smith and Elliott (1878), and Turrill (1876) all compare the valley to a park. Bartlett (1854), Nordhoff (1873), Cone (1876), and Whitney (in Turrill 1876) referred to a landscape gardener or designer.

[22] Bartlett (1854). See also Avery (1878).

[23] Stewart (2002), Anderson (2005: 174–179), and Lightfoot and Parrish (2009).

[24] Altimira (1861: 61) and Smilie (1975).

[25] Local ethnographies (e.g., Driver 1936) identify some of these reasons for burning; other reasons are consistently documented in other California tribes (e.g., Stewart 2002 and Anderson 2005).

[26] Smilie (1975: 61).

[27] Driver (1936). Driver's study focused on the Wappo-speaking people of the Alexander Valley in Sonoma County; the degree to which terminology in the Napa Valley was similar is not known (Milliken 1978: 2.7–2.8).

[28] Jepson (1909) referred to the savannas as "oak orchards," stating that the "extent and nature of the relation of Indian tribal culture and the habit of the oaks cannot yet, if ever, be completely defined, although it is clear that the singular spacing of the trees is a result of the annual firing of the country—an aboriginal practice of which there is ample historical evidence." Indigenous fire management was not limited to the valley, but was also practiced in the adjacent hills to control the expansion of woody vegetation. Burning was continued by settlers, at least to some degree, to maintain hunting areas (Grossinger et al. 2004b).

[29] Altimira (1861).

[30] Jepson (1910), Pavlik et al. (1991), and Brown (2005). Witness trees (also called "bearing" trees) were selected to establish township and range section corners. Each tree's species, diameter, azimuth, and distance from the survey point was recorded in the field notes of the survey.

31 White oak: *Daily Alta California* (1860), Hittell (1863), and Cronise (1868). Burr oak: Turrill (1876) and Cone (1876). Willow-oak: Avery (1878); "willow-oak" was presumably related to the term "weeping oak," referring to the pendulous, dangling branches of mature trees (Jepson 1910). Valley oak: *California Star* (1848). The term *roblar* (white oak grove) was also used in Spanish *diseños*.

32 Jepson (1912) and Carpenter and Cosby (1938: 3) affirmed the presence of black oak in the valley. Brace (1869) also noted *Q. sonomensis*, the former name for black oak, although he referred to the tree as evergreen.

33 Bartlett (1854). Brace (1869) and Avery (1878) both mentioned that there were several species of evergreen oaks. Coast live oak *(Q. agrifolia)* is most common on the valley floor today, but interior live oak *(Q. wislizeni)* could also have been present.

34 Bowles (1865), Evans (1873), and *San Francisco Call* (1895a). Jepson considered coast live oak "fairly common" in 1912.

35 Cone (1876).

36 Bartlett (1854) and *Daily Alta California* (1864). In addition, Menefee (1873) described "occasional gigantic madronas" within the oak savannas. The suggestion by Cronise (1868) that sycamore was a significant component of the savannas is not confirmed.

37 Jepson (1912), *California Star* (1848a, b), Taylor (1862), Bartlett (1854), and Carpenter and Cosby (1938). Carpenter and Cosby's statement that "in the valley are still growing a few clumps of valley oak, black oak, live oak, and Digger [gray] pine which once covered the valley floor" closely matches the GLO and textual evidence.

38 Avery (1878).

39 Avery (1878).

40 See Hamilton (1997) and Holstein (2001) for discussion of the "bunchgrass hypothesis." See Shiffman (2007), Minnich (2008), and Sawyer et al. (2009) for discussion of wildflower evidence. Rhizomatous ryegrasses *(Leymus* spp.) were also likely a major component (Holstein 2000 and 2001).

41 Bartlett (1854).

42 Caton (1879).

43 George Yount's epiphany vision of the valley to which he would settle was that "it was gay with early *eschscholtzia*" *(Eschscholzia californica,* California poppy), according to his granddaughter (Bucknall 1917).

44 Jepson (1912). Jepson (1893 and 1897) also collected white forget-me-nots (rusty-haired popcorn flower, *Plagiobothrys nothofulvus*) repeatedly between St. Helena and Rutherford, illustrating another component of the flower fields. The fragrant species (called *nievitas* in Spanish, the diminutive of *nieve,* snow; Parsons 1909) often covers entire fields, giving the appearance of a light snowfall.

45 Minnich (2008).

46 Grossinger et al. (2008) and Minnich (2008).

47 Ingols (2006: 10).

48 Hittell (1863).

49 Barlett (1854) and Cronise (1868).

50 Evans (1873).

51 Hittell (1863). See also *Daily Alta California* (1860) and Brace (1869).

52 Researchers speculate that humans evolved in these heterogeneous environments with sufficient open space for hunting and travel, as well as trees for cover and shelter (e.g., Wilson 1984: 109; Bock and Bock 2000: 70–71).

53 *Sacramento Daily Union* (1860).

⁵⁴ Smith and Elliott (1878) and Babe Learned, interview by Elise Brewster, October 5, 2002.

⁵⁵ Smith and Elliot (1878).

⁵⁶ Densities of canopy trees that were sampled in nine remnant areas ranged from 0.3 to 0.8 trees per acre, for an overall density of 0.5 (854 trees in 1,677 acres; see also note 17). We used this value for the early 1800s estimate. Given the uncertainty in the area and density, we classify this estimate as medium certainty (+/- 50%).

⁵⁷ Jepson (1909).

⁵⁸ Geiger (2004), McPherson (2007), and Xiao et al. (2000).

⁵⁹ Manning et al. (2006), Grossinger and Beller (2011), Tietje (2011), and Whipple et al. (2011). Important elements for landscape-level ecological function include links to upland and stream habitats, granary trees (for acorn woodpeckers), standing snags and dead limbs (for cavity-nesting birds and foraging on insects), untended fallen branch litter under certain trees (for foraging by Pacific pallid bat). Much research has been done on design and maintenance strategies for large trees in urban areas (Costello et al. 2011).

⁶⁰ Sork et al. (2002).

3 • CREEKS

¹ Menefee (1873: 148). The quote refers to Bear Canyon Creek near Rutherford.

² Leidy et al. (2005a) and Becker et al. (2007).

³ Menefee (1873: 92, 147).

⁴ For a discussion of discontinuous streams in other parts of the Bay Area, see Sowers and Thompson (2005), Grossinger et al. (2006 and 2008), and Beller et al. (2010). Benner and Sedell (1997) discussed similar patterns in the Willamette Valley, Oregon.

⁵ Knighton (1998).

⁶ Carpenter and Cosby described the phenomenon of the Napa River's natural levee: "In the central part of the valley, the Napa River occupies a comparatively deep channel but frequently overflows its banks and builds up a channel ridge that shuts off surface drainage from the bordering flood plain." Similar natural main stem barriers to tributary drainage have been observed on other coastal California streams such as the Salinas River (Beller and Grossinger, unpublished research) and Coyote Creek (Grossinger et al. 2006).

⁷ Coombs (1861).

⁸ Carpenter and Cosby (1938: 65).

⁹ Menefee (1873: 132).

¹⁰ Carpenter and Cosby (1938: 49, 51, 54) compared the effect of a high water table on crops to that of a hardpan, explaining: "By drowning the roots that penetrate the lower subsoil during the dry season, a fluctuating water table just as effectively limits the rooting zone of plants as if the soil were underlaid at a slight depth by an impervious substratum or bedrock. The result is that the rooting system is confined largely to the surface soil and the plant suffers from lack of moisture during the dry season" (65).

¹¹ Carpenter and Cosby (1938: 65).

¹² CDWR (1995) and NCFCWCD (2005). In contrast to the larger Napa Valley groundwater basin, the isolated Carneros and Milliken-Tulucay-Sarco basins have experienced significant long-term decline.

¹³ NCFCWCD (2005).

¹⁴ SFEI (2011a).

NOTES TO PAGES 55-64 • 187

¹⁵ Descriptions of dry farming in the Napa Valley are quite consistent through the 19th and 20th centuries. Smith and Elliott (1878: 18) reported that "fruit in Napa is mostly grown in the valleys, where the land is so low that sufficient moisture can be obtained by plowing the land two or three times during the season." Bryan (1932: 3) observed that "the present use of water for irrigation in Napa Valley is very small and successful agriculture and horticulture is pursued without irrigation." Carpenter and Cosby (1938: 65) concluded that "with a water table stabilized below a depth of 6 feet, little or no irrigation of most deep-rooted crops grown in the valley would be necessary." Kunkel and Upson (1960: 8–9) noted that "the alluvial plains [of Napa and Sonoma valleys] are ideally suited to dry farming of grapes and prunes." It is unclear to what extent such conditions remain, given the potential effects of soil compaction and subsurface drainage. However, dry farming is still done in places, and the practice of limiting supplemental water in order to maximize grape flavor (deficit irrigation) has increased in recent years, suggesting the existing potential of the natural sponge effect to support agriculture.

¹⁶ For more information about model calibration and results, see SFEI (2011a).

¹⁷ SFEI (2011b).

¹⁸ Berner et al. (2003) and Bousman (2007).

¹⁹ Baird (1870) and Grossinger et al. (2004b). This trend is consistent with Leopold's (1994) observation that the spontaneous return of riparian vegetation has helped initiate a "state of healing" on many streams in the western United States since the 1950s.

²⁰ Leach (1917). The creek curved directly around Napa Hotel in downtown Napa before its confluence with the Napa River.

²¹ Scott et al. (1996) and Knighton (1998).

²² Beagle (2010).

²³ Mount (1995).

²⁴ Pearce and Grossinger (2004). Levee building, gravel mining, removal of snags, intensive grazing, soil compaction, and watershed deforestation can all also cause incision. For more discussion, see Cooke and Reeves (1976), Dunne and Leopold (1978), and Mount (1995).

²⁵ Simon and Rinaldi (2006).

²⁶ Micheli and Kirchner (2002) and Simon and Rinaldi (2006).

²⁷ Beagle (2010).

²⁸ Beagle (2010).

²⁹ Carpenter and Cosby (1938).

³⁰ Kondolf (2006).

³¹ Sapozhnikov and Foufoula-Georgiou (1999).

³² Sulphur Creek: Grossinger et al. (2004b); Conn Creek: Weber (2001: 248).

³³ USGS (1902).

³⁴ Canyon reaches often provided more reliable summer water. Menefee (1873: 47, 48, 92) and Smith and Elliott (1878) describe the use of "mountain streams" diverted near the canyon mouth to supply residential uses and gardens.

³⁵ Wallace (1901).

³⁶ Sanborn Perris Map Company (1886).

³⁷ Vines (1861: 295–297).

³⁸ Altimira, in Smilie (1975: 8).

³⁹ Grossinger et al. (2004a).

40 Grossinger et al. (2004a). The expansion of the terrestrial vegetation in the upper watersheds of these creeks due to the cessation of indigenous burning may also contribute to reduced base flow, through increased transpiration (Grossinger et al. 2004a and Anderson 2005: 156).

41 NCFCWCD (2005).

42 Stillwater Sciences and Dietrich (2002).

4 • VALLEY WETLANDS

1 Brierley et al. (1999), Walter and Merritts (2008), and Fryirs (2010).

2 CDWR (1995).

3 Winfrey (1953). The blackberry patches were part of the Bale Slough wetland complex.

4 Shalowitz (1964: 201).

5 USGS (1913).

6 Goals Project (1999) and Hulse et al. (2002).

7 We combined appropriate soils from both Carpenter and Cosby (1938) and Lambert and Kashiwagi (1978) to define the historical extent of wet meadow. Given that farmers began to drain low-lying parts of the valley as early as the 1850s (Menefee 1873), the extent of wet meadows indicated by these 20th-century sources is probably conservative. The choice of soil types was calibrated against the earlier but more generalized Holmes and Nelson (1917) survey and the distribution of known historical wetlands. We refined the potential wet meadow areas to exclude a few areas where we had strong evidence for the historical presence of oak savanna, based upon previous findings that oak savanna is strongly non-coincident with wet meadows (Grossinger et al. 2006). We also adjusted the wet meadow areas to conform to more precise boundaries of tidal marshland and uplands (hills). Soil types with some potential wet meadow characteristics (e.g., claypan) in the Carneros area were excluded because they occur on sloping terraces and are not recent historical depositional environments. Soil type boundaries are inherently limited in precision, as one type naturally intergrades into another. As a result, the meadow edges should be considered approximate. Additionally, mapped areas typically contain smaller intrusions that are impractical to map as separate features, so wet meadows may contain minor, well-drained areas within their boundaries.

8 Carpenter and Cosby (1938: 52–53, 59–60).

9 Jepson (1912: 152). Jepson (1936a: 16–17) also collected in "the low wet hollows" of the Napa Valley. Near Rutherford, perhaps in the Bale Slough wetland complex, he described freshwater marsh and wet meadow associates such as watercress and basket sedge.

10 Michener and Bioletti (1891).

11 Carpenter and Cosby (1938: 48–49).

12 Records from the Consortium of California Herbaria at http://ucjeps.berkeley.edu/consortium.

13 Ruygt (2001).

14 Holmes and Nelson (1917: 97).

15 Holstein (2000) and Faber (2005).

16 Jepson (1918) and Hall (1923). Carpenter and Cosby (1938: 63–64) also reported that "saltgrass, locally known as 'devil grass,'" was found in salt-affected areas within the valley, suggesting additional areas of alkali meadow.

17 Ruygt (1982a, 1982b, and 1991). Data provided by the Consortium of California Herbaria at http://ucjeps.berkeley.edu/consortium.

18 Cooper (1926), Baye et al. (2000), Holstein (2000), and Grossinger et al. (2006).

19 SFEI (2011b).

20 To identify remnant wetlands, we selected contemporary wetlands (outside of the baylands) that fully or partially intersected with the location of historical wetlands. We then reviewed each potential remnant to remove those clearly associated with constructed features (e.g., wetlands within ditches or reservoirs).

5 • NAPA RIVER

1 Skinner (1962) and Leidy (2007).

2 Tallies vary, but generally include 1852, 1861, 1870, 1878, 1879, 1880–81, 1889, 1892, 1934, 1940, 1941, 1950, 1952, 1955, 1958, 1963, 1964, 1986, 1991, 1995, 1997, and 2005 (Nolte 1959 and Dillon n.d., 2004: 12).

3 Wichels (n.d.: 15).

4 Rick Thomasser, NCFCWCD, pers. comm.; modeling by SFEI (2011a) also suggests increases in peak flow.

5 Nolte (1959) and Ketteringham (1961).

6 Smith and Elliott (1878).

7 By the 1920s, river pollution in the vicinity of Napa was so severe that it was only possible to "catch fish in town for a few weeks out of the year," when the rains had "purified the river" (Scofield and Bryant 1926).

8 The Rutherford reach project, led by the Napa River Rutherford Dust Restoration Team, focuses on Zinfandel Lane to Oakville Road. The contiguous Yountville reach project, by the California Land Stewardship Institute and local partners, continues downstream to Oak Knoll Road (CLSI 2011).

9 Camp (1966), Clyman (1960), USDC (ca. 1840b), and USDC (1841).

10 Squibb (1861). The river divided the Yahome and Napa ranchos; Squibb was testifying in the Yahome case.

11 Gray (1853).

12 Henning et al. (2006).

13 Adams (1946: 48).

14 Also known as the main, primary, or low-flow channel. The main channel carries most of the flow at high river stages and all of the flow at low stages.

15 Historical slough widths recorded by GLO Public Land Surveys range from 15 to 75 links wide, translating to 10–50 feet (100 links equals 1 chain, or 66 feet):

> "Slough from creek" 50 links wide in downtown Calistoga (Tracy 1858b: 51).
>
> Sloughs 30 links wide south of Calistoga (Dewoody 1867: 202)
>
> Sloughs 50, 20, 30, and 20 links wide in willows near Zinfandel Lane (Tracy 1858c: 811)
>
> Sloughs 30, 50, 15, 50, 30, 75, 75, and 50 links wide (Tracy 1858a: 198–201).

16 Vallejo (1861: 238-245), Coombs (1861: 559), and Barnett (1861: 125). The Caymus land case is a good example of the value and limitations of court testimony. In this case, while testimony about the names and ownership of specific islands is confusing and potentially contradictory, the general concept of multiple islands in this part of the valley is supported by all witnesses.

17 Coombs (1861: 545-548).

18 Bickford (1928 and 1931). Grinnell et al. (1918: 145) confirmed the species' presence at the turn of the century: "Mr. A. Jackson reports that a limit of Wood Ducks could often be obtained fifteen years ago along the Napa River, but that now not a single Wood Duck is to be seen there."

[19] Sedell and Luchessa (1982), Grinnell et al. (1937), and Pollock et al. (2007).

[20] Length of active side channels, 32 miles; main stem Napa River, 36 miles. Measurements based on the historical Napa Valley GIS.

[21] Sommer et al. (2001), Boughton et al.(2006), Henning et al. (2006), Moyle et al. (2007), and Jeffres et al. (2008), McBain and Trush (2008), Cluer et al. (2009).

[22] Nolte (1959). Because of its small scale (11 x 17 inches), the map, while quite detailed, exhibited some spatial inaccuracy when georeferenced. We edited the map to conform to river alignment and topography; these adjustments slightly (less than 3%) reduced the total area of flooding. Accordingly, flood boundaries should not be considered precise and may include spatial error of 1,500 feet or more.

[23] Moyle et al. (2007).

[24] Gillette (1972) and Wohl (2004).

[25] Detailed maps showing the river before we would expect substantial modifications include Tracy (1858d), Dewoody (1870 and 1873), Morgan (1871), and USSG (1869).

[26] Kondolf (2006).

[27] Sinuosity is a measure of a stream's "curviness": a value of 1 would be perfectly straight; 1.5 is considered the minimum for a meandering (Mount 1995 and Beechie et al. 2006) or a highly sinuous (Rosgen 2007) stream. Part of the difference between the Napa River and more serpentine rivers such as the Sacramento is that the Napa Valley is comparatively steep (as discussed on page 5).

[28] Schumm (1985), Beechie et al. (2006), and Rosgen (2007).

[29] Beechie et al. (2006) and Rosgen (1994 and 2007). Sinuosity values represent the stream length divided by the corresponding valley length. For individual reaches, valley length was measured as a single straight line segment. For the full river measurement, valley length was composed of three segments to follow the alignment of the valley/meander belt (Mount 1995).

[30] Cooke and Reeves (1976) and Trimble (2003).

[31] SFEI (2011a).

[32] Fisher (1959).

[33] Stillwater Sciences and Dietrich (2002) estimated 6–8 feet of incision; see also Napolitano et al. (2009). Comparison of longitudinal profiles were based on USGS topographic quadrangles (circa 1950s data). Cross sections done by the Napa County RCD in 1996 also corroborate 10–13 feet of incision downstream of approximately St. Helena, and lesser amounts upstream (SFEI 2011a)

[34] Napa County (1921).

[35] Stillwater Sciences and Dietrich (2002) and Napolitano et al. (2009).

[36] Prominent unvegetated gravel or sand deposits are consistently present at sites along the river in aerial photograph sequences from the 1940s through the 1960s, suggesting that they were not anomalous effects of recent floods or manual clearing. These features are also affirmed by accounts of mining from prominent bars to supply building materials. For example, Weaver (1949) identified a significant site 2 3/4 miles north of Napa: "gravel, sand, and loam are obtained by selecting material from bars along a half-mile section of the river"; see also Weber (1998: 194). In the 1950s, as plans for widening the river were prepared, the Department of Fish and Game requested that one side of the river be retained "in its natural state as this serves as a spawning ground for Steelhead and the removal of the shade for this bank would endanger this important fishery resource" (Nolte 1959).

[37] Stillwater Sciences and Dietrich (2002) and Napolitano et al. (2009).

[38] Napolitano et al. (2009) and SFEI (2011a). These changes to stream function were anticipated prior to reservoir construction. Mary Grigsby, who operated a gravel mining operation on lower

Conn Creek, argued against the construction of Conn Creek Reservoir in the 1940s, anticipating the loss of annual gravel replenishment (Weber 2001: 248).

[39] *San Francisco Call* (1895b).

[40] Mendell (1885); also reported in the *San Francisco Chronicle* on May 13, 1888.

[41] Much research and policy has been developed recently to address sediment challenges in the river. See Stillwater Sciences and Dietrich (2002: 38), Napolitano et al. (2009), CLSI (2011) and SFEI (2011a).

[42] Grossinger et al. (2006 and 2008).

[43] Jonathan Koehler, pers. comm. Alicia Gilbreath and Lester McKee (unpublished data), in an analysis of USGS flow records at Oak Knoll, found 30-day periods of low or no flow (less than 1 cfs) to be common in recent decades.

[44] Stromberg and Patten (1992), Stromberg et al. (2005), and White and Greer (2006).

[45] McGlashan and Dean (1913), Bryan (1932), and Faye (1973).

[46] Thompson (1857: 49).

[47] Vallejo (1885), Tortorolo (1978), and Sanborn (1910). A newspaper at the time reported that "a large force of men are at work, connecting the system with the boundless supply of the Napa river" *(Pacific Rural Press* 1885).

[48] McGlashan and Dean (1913).

[49] McGlashan and Dean (1913).

[50] Joseph Grinnell, on November 28, 1936, described an "afternoon driving back and forth along 'lanes' on the floor of Napa Valley, reaching Calistoga" from Pacific Union College. The preceding water years 1934–35 and 1935–36 were within 2% of average.

[51] Tom Wilson, interview by Shari Gardner on March 22 and 29, 2001, and March 4, 2003; Ingols (2006) refers to "the Wilson Pool."

[52] In an assessment of flow needs for fisheries, CDFG (ca. 1979) noted that "the cumulative, unregulated demand for water is so great that it appears possible for even winter flows to be entirely diverted in some years." Kunkel and Upson (1960) considered the river to be intermittent for much of its length. The presence of beaver ponds would also have contributed to increased dry season flow (Tappe 1942 and Pollock et al. 2007).

[53] Bryan (1932).

[54] Leidy (2007) and NCFCWCD (2005).

[55] Based on St. Helena long-term rainfall average, in Grossinger et al. (2004b). Table after McGlashan and Dean (1913).

[56] See Thomas (2004) and Collins et al. (2006) for summaries of riparian buffer widths required for different stream functions. Physical processes such as sediment entrapment and pollution filtration can be accomplished by relatively narrow riparian zones, but most wildlife species require more space and some need very wide areas. For example, research on the yellow-billed cuckoo suggests that sites narrower than 100 meters (325 feet) are unsuitable for breeding (Laymon 1998 and Laymon and Halterman 1989).

[57] Dewoody (1873).

[58] Belden (1887) and Grossinger et al. (2008: 26–27).

[59] Tracy (1859), Smith and Elliott (1878: 3), Gregory (1912: 882), Winfrey (1953), Weber (1998: 123–24), and Stone (2003). George Yount's granddaughter Mary Bucknall (1917) also described a "willow copse" that was likely located along the sloughs near Yount's Mill.

[60] *Pacific Rural Press* (1871).

61 GLO surveys showed that the river near Zinfandel Lane was half as wide as reaches immediately upstream and downstream; the system's capacity was expanded, however, by the four adjoining sloughs averaging 20 feet wide. Upstream and downstream of the wetlands, the GLO recorded the river as 1.5–2 chains wide (99–132 feet); similar width is observed in 1940 aerial photography. In contrast, the two surveys that crossed the wetland complex described the channel as 0.75–1 chains wide (50–66 feet; Thompson 1857 and Dewoody 1866b).

62 SFEI (2011b)

63 This approach does not produce a precise measurement, but rather a general estimate of the relative amount of riparian forest of differing widths. Riparian forests of these widths are not unprecedented in the region. Beller et al. (2010) documented willow-dominated riparian forests 500 to 1,500 feet wide (on one side) at similar elevations along Guadalupe River (Santa Clara County), a stream of similar size in a valley with high groundwater. These widths are small compared to some larger, humid-climate rivers of the western U.S. such as the Willamette River (1.5–3.5 kilometers on a side; Sedell and Froggatt 1984) and rivers of the Puget Lowland (Collins et al. 2003).

64 Baird (1870), Grayson (1986), Berner et al. (2003), and Bousman (2007).

65 Grinnell et al. (1937).

66 Work (1832–33), in Maloney (1943: 43–44, 55, 94–95); careful reading of the original April 9 entry also suggests that it may more accurately refer to a tributary to the Napa River than to the whole watershed. Skinner (1962: 162) also reported that beaver were historically present on the Napa River. The river's extensive side channels may have provided particularly good habitat, where dams were less likely to be blown out by high flows (Grinnell et al. 1937 and Pollock et al. 2004).

67 Naiman et al. (1988) and Pollock et al. (2007).

68 The value of beaver to salmon restoration and the recovery of incised channels is well documented in the Pacific Northwest (e.g., Pollock et al. 2004, Saldi-Caromile et al. 2004, and Beechie et al. 2006). Less research has been done in California, but Tappe (1942) described the importance of beaver ponds to trout populations in streams that otherwise would be dry (including a tributary to Putah Creek in Napa County).

69 Jonathan Koehler, pers. comm.

70 Farmers have preserved beaver colonies for their dams, which can serve as natural reservoirs, raise the local water table, and maintain summer base flow (Grinnell et al. 1937: 659, 715–16 and Tappe 1942).

71 Stillwater Sciences and Dietrich (2002).

72 Sedell and Froggatt (1984), Triska (1984), and Collins et al. (2002).

73 Cobb (1858), Mendell (1885), and Dillon (2004: 122).

74 The *St. Helena Star* reported a major logjam in 1875, apparently in the vicinity of St. Helena (Dillon 2004: 122). Dillon (2004: 11) also described an 1878 flood that uprooted and carried trees 5 feet in diameter downstream. Anecdotal accounts recollect the common removal of logs to reduce the potential for flooding (Shari Gardner, pers. comm.). Conifers supplied from the upper watershed by well-connected tributaries (e.g., Sulphur Creek) may have been important sources of durable wood jam pieces (Collins and Montgomery 2002).

75 Harwood and Brown (1993), Collins and Montgomery (2002), Collins et al. (2002), and Sear et al. (2010).

76 Jepson (1892: 110, 1900: 47, 1900: 67, 1903: 106, 1922: 141, and 1934: 66).

77 Jepson (1934: 67).

78 Jepson (1912 and 1936b).

79 Jepson (1910 and 1912: 151).

80 Revere (1849) and Thompson (1857: 478).

81 D'Hémécourt (1855), Squibb (1856), Thompson (1857: 464), Tracy (1858a), Sonne (1897), and Rees (1914). Spring and Lewis (1932) and Avery (1878) (sycamore reference may be a mistake or refer to a rare constituent).

82 Ruygt (unpublished data).

83 It should be considered whether riparian tree species might have been preferentially cleared for timber or firewood. However, riparian vegetation was not of particularly high value. Describing the resources between Yountville and Rutherford, Thompson reported that "on the banks of the creek and the slough there is an abundance of timber of an inferior quality for all purposes of fuel" (Thompson 1857: 489). Cottonwoods, sycamore, and oaks provided poor building materials (Jepson 1910). Some areas may have been impacted by firewood harvest, but oak firewood was available throughout the valley and there is no evidence for large-scale riparian clearing.

84 Sycamore was the most commonly reported tree along much of Coyote Creek and other streams in Santa Clara County (Grossinger et al. 2006, 2007, and 2008).

85 Cottonwood expansion into formerly more open channels has been observed on other California streams (Grossinger et al. 2007 and 2008; Dilts forthcoming).

86 Menefee (1873: 12) described the vegetation of Napa County riparian corridors: "All [the valleys in the county] are intersected by water courses, whose sinuous banks are fringed with trees and shrubs. Laurel, live oaks, buckeyes, manzanitas, alders, willows and the ash, are the principal trees. Of shrubbery there is a great variety, among which we name ceanothus or California lilac, elder, bay, and hazel nut. There are also in many places, large tracts covered with a species of dwarf holly, bearing beautiful red berries in heavy clusters [presumably toyon]. Wild grape vines abound along every stream, and used to afford the grizzly a considerable portion of his provisions during the fall." Five years later, the description was copied nearly verbatim by Smith and Elliott (1878: 3): "All the valleys are traversed by water courses whose banks are fringed with trees and shrubs. Laurel, live oaks, buckeyes, manzanitas, alders, willows, and the ash, are the principal trees."

87 Le Conte (1885) and *Sacramento Daily Union* (1891). See also Swett and Aitken (1975: 11) for a photograph of this riparian structure.

88 Leidy (2007) and Koehler (unpublished data).

89 Koehler and Blank (2010).

90 Leidy et al. (2005a, b), Stillwater Sciences (2006), Leidy (2007), and Jonathan Koehler, pers. comm.

91 NCALM (2003).

92 Schumm et al. (1984); Andy Collison, pers. comm., based on fieldwork in the Oakville–Oak Knoll reach; Napolitano et al. (2009); and CLSI (2011).

6 • TIDAL MARSHLAND

1 Mitsch and Gosselink (2000).

2 Goals Project (1999). A guidebook to California hunting reported that the Napa River's marshes provided "feeding grounds for water fowl as well as for snipe, rail and larks, with quail on the nearby uplands" (Southern Pacific Company 1896).

3 Metrics in this paragraph from Goals Project (1999). See Table 1 for quantification of regional changes in bay habitats.

4 While detailed mapping for the full coastline is still not available, Atwater and Hedel (1976) estimated that three-quarters of California's coastal wetlands were found in the San Francisco Estuary. The area of the Napa marshes is defined as the tidal marshes with primary tidal connection to the Napa River (west to Napa Slough, downstream to Mare Island), measured

from the Bay Area EcoAtlas. As Gilbert (1917) noted, this is a somewhat arbitrary boundary to divide the Napa River's marshes from those draining to Sonoma Creek. However, he divided them similarly, estimating 26 miles (16,640 acres) of marsh associated with the Napa River, without the benefit of detailed T-sheet reproductions and GIS.

[5] As part of his seminal study of the impact of hydraulic mining debris, Gilbert (1917: 103) studied the effect of marsh reclamation on the shoaling of bay channels. He predicted that reclamation would soon "exclude the tides from all the marsh area except some of the larger sloughs" and that this would have the effect of reducing tidal discharge—and associated main channel volume—by about two-thirds.

[6] This style was a variation on the official symbol, which should have included vegetation "tufts" (Shalowitz 1964: 191).

[7] Mature tidal marshes lie above normal high tide (Goals Project 1999); based on their complex drainage patterns and elevation, Napa River's tidal marshes appear to have been mostly high-elevation, mature systems. *Pacific Rural Press* (1875) stated that they "are quite above the ordinary flood tides in Summer. They are only submerged from four to six inches by the excessive high tides, which occur only occasionally." Stanly (1885) might have had some lower elevation marshes on his property: "Before reclamation all the land was covered by the spring tides, and a very large part by every high tide."

[8] USDC (1840c), Taylor (1862), *Pacific Rural Press* (1875), Menefee (1873: 13, 81, 138-39), and Wallace (1901).

[9] Clyman (1960).

[10] Brewer (1966). The tidal marshes north of the Suscol Narrows, which had large ponds and relatively few tidal channels, had characteristics associated with freshwater tidal systems (Grossinger 1995).

[11] The term *estero* is used rather than *tular*, for example, in *diseños* showing the tidal marshlands along Coyote Creek, and mentions of tules are restricted to limited freshwater-influenced zones (Grossinger et al. 2006).

[12] *Pacific Rural Press* (1885).

[13] Walter Carvelli, interview by Josh Collins on March 21, 1994.

[14] *Pacific Rural Press* (1875).

[15] Carpenter and Cosby (1938). Similarly, County Surveyor Dewoody (1860) described tidal marsh plains between Sonoma and Huichica creeks (which had less direct freshwater influence from the river) as having a much smaller proportion of tules: "The growth at present on the marsh consists of a variety of salt grass, weeds, and a few bushes, and on the banks of some of the sloughs clover and pea vines, and in some places tules." Stanly (1883) stated that before reclamation his land "was covered by a dense growth of tules upon its lower levels and by a tough sod of salt or wire grass upon the higher ground."

[16] Grinnell (1901), Atwater and Belknap (1980), Knight (1980), and Stanford et al. (forthcoming).

[17] Records from the California Natural Diversity Database of the California Department of Fish and Game.

[18] Carpenter and Cosby (1938).

[19] Stanly (1883) described how this fortuitous effect "gives us fresh water in Napa river, as long as there is a large body of water flowing down the Sacramento and San Joaquin rivers." Warner (2000) reported a similar hydraulic effect at Mare Island that resulted in net sediment transport upstream from Carquinez Straits. Salinity patterns are highly variable and thus challenging to compare through time. However, recent monitoring at the South Wetlands Opportunity Area (just upstream of Stanly Lane and near the location of historical accounts) does show that surface salinity typically exceeds 7 parts per trillion by late July, even in relatively wet years—too salty for irrigation (Stillwater Sciences 2006: Figure A-8, Table C-1).

[20] Bar formation was ascribed to floods occurring every two to three years, for which tidal currents were "ineffectual to their removal" (Mendell 1885 and Rees 1914).

[21] Mendell (1885).

[22] Suscol may have been the better natural site for a port. Historian Hittell (1863) commented on the problematic location of the town of Napa, which has necessitated over a century of dredging: "If mere natural advantages were to be taken into account, the town [Napa] would be at Suscol, which is six miles nearer to the bay, and always accessible by small steamers, while at low tide the boats must stop several miles below Napa."

[23] Alden (1860) and Mendell (1885).

[24] Le Conte (1885).

[25] Rees (1914).

[26] Dillon (2004: 122).

[27] Menefee (1873) and Smith and Elliott (1878). Similar promise was described just to the south, where, after constructing levees, Mr. V. Hathaway was able to gather "a most remarkable crop of plums…from trees growing upon this piece of Swamp Land, nothing more being required than a small load of earth to each tree, dumped upon the marsh, and the tree set in it" (Houghton 1862).

[28] *Pacific Rural Press* (1885).

[29] Skinner (1962: 72).

[30] Byrns (1922).

[31] Corps of Engineers (1942).

[32] Carpenter and Cosby (1938): 54–55 and Madrone Associates (1977). Biologist Joseph Grinnell observed duck hunting in the recently reclaimed marshes: "Along this road…with the great Napa tidal marshes (now much reclaimed interiorwards), we saw much of interest and tarried frequently. This was next to the last day of the duck-shooting season, and even tho it was late in the day, we heard many shots, from the duck-clubs on the marsh inland and from out on the Bay" (Grinnell 1936).

[33] By the 1920s, "the mosquito problem had become unbearable and as a detriment to use and livability was realized to be a serious factor in the future development" of the southern part of the Napa Valley, affecting areas from Napa to Mare Island (NCMAD 1934 and Whitthorne 1969). As they dried, the peat soils formed giant cracks that had to be filled by plowing. A mosquito control report of the era described the "dead sloughs and ponds" of the now-diked marshlands as primary breeding areas.

[34] Map of 1978 tidal marsh extent based on USGS quadrangles of Cuttings Wharf (1981), Napa (1980b), and Mare Island (1980a), all of which used 1978 photography. Contemporary map based on SFEI (2011b) and the California Wetlands Portal (www.californiawetlands.net). The 1978 view is more approximate than the other two because of less detailed original mapping. The contemporary map includes mapping from the period 2007–2009; projected future totals will likely change as plans evolve.

[35] Lydia M. Money, interview by Robin Grossinger on April 6, 2009. More recent publications have dated the marsh reversion to 1980, but the Money family and Madrone Associates (1977) confirmed that the levee breached in the 1950s. Bull Island is full of odd stories: water for irrigation was obtained through a surplus firehose running all the way across the Napa River to a water tank in the Carneros; the island was supposedly known for its unusual long-legged "tule rabbit," adapted to see over the tall marsh vegetation: "the silliest-looking rabbit you ever saw" (interview with Lydia Money).

[36] Goals Project (1999: A-16) and Baye et al. (2000).

[37] Collins et al. (2007).

[38] Goals Project (1999: 14–20, 107); BCDC (2009: 146–149), and Heberger et al. (2009).

[39] Duhig (1990).

[40] Stevenson (1884).

[41] Knowles (2010).

7 • LANDSCAPE TRANSFORMATION AND RESILIENCE

[1] Swetnam et al. (1999), Collins and Montgomery (2001), Collins et al. (2003), and Simenstad et al. (2006).

[2] Hamilton (1997), Kondolf et al. (2001), Foster and Motzkin (2003), and Montgomery (2008).

[3] Sedell et al. (1990) and Leidy et al. (2011).

[4] McBain and Trush (2008), Waples et al. (2009).

8 • LANDSCAPE TOURS

[1] See www.californiawetlands.net/tracker/ba/map.

[2] Stevenson (1884).

[3] Records from the California Natural Diversity Database of the California Department of Fish and Game.

[4] Gardner (1977).

[5] Menefee (1873), Soderholm (n.d.), Gunn and Hunt (1926), and Gardner (1977).

[6] *Daily Alta California* (1860).

[7] Duhig (1990).

[8] Knowles (2010).

[9] Brewer in 1861, imagining the Napa Valley at Yount's initial visit two decades earlier (Brewer 1966).

[10] Leach (1917).

[11] Sanborn Perris Map Company (1886: 3 and 1891: 9).

[12] Sanborn Perris Map Company (1891: 9) and Sanborn Map Company (1910: 14).

[13] Tortorolo (1978) and Grossinger and Beller (2007).

[14] Wichels (n.d.: 22–23).

[15] Bucknall (1917).

[16] Wichels (n.d.).

[17] Berner et al. (2003).

[18] Menefee (1873: 14).

[19] Camp (1966: 159–161).

[20] Camp (1966: 159).

[21] Smith and Elliot (1878: 15).

[22] Wichels (ca. 1976).

[23] Adams (1946: 48).

BIBLIOGRAPHY

Adams, I. C. 1946. *Memoirs and anecdotes of early days in Calistoga*. [Calistoga, CA]: Privately Printed. Courtesy of The Bancroft Library, UC Berkeley.

Alden, J. 1860. *Hydrography of Napa Creek, California, register no. H723*. Washington, DC: U.S. Coast Survey.

Allen-Diaz, B., J. W. Bartolome, and M. P. McClaran. 1999. "California oak savanna." In *Savannas, barrens, and rock outcrop plant communities of North America*, eds. R. C. Anderson, J. S. Fralish, and J. M. Baskin, 322–339. Cambridge: Cambridge University Press.

Altimira, J. 1861. "Journal of a mission founding expedition north of San Francisco in 1823." In *Hutchings California Magazine, July 1860-June 1861*. San Francisco: Hutchings & Rosenfeld. Courtesy of Sonoma County Library.

Anderson, K. 2005. *Tending the wild: Native American knowledge and the management of California's natural resources*. Berkeley, CA: University of California Press.

Atwater, B. F. and D. F. Belknap. 1980. "Tidal-wetland deposits of the Sacramento–San Joaquin Delta, California." In *Quaternary depositional environments of the Pacific Coast: Pacific Coast Paleogeography Symposium 4*, eds. M.E. Field, A.H. Bouma, D. Colburn, and Ingle. Los Angeles, CA: The Pacific Section Society of Economic Paleontologists and Mineralogists.

Atwater, B. F. and C. W. Hedel. 1976. *Distribution of seed plants with respect to tide levels and water salinity in the natural tidal marshes of the northern San Francisco Bay estuary, California*. Open-file report 76-389. Menlo Park, CA: U.S. Geological Survey.

Avery, B. P. 1878. *Californian pictures in prose and verse*. New York: Hurd and Houghton.

Bahre, C. J. 1991. *A legacy of change: Historic human impact on vegetation in the Arizona borderlands*. Tucson: University of Arizona Press.

Balée, W. 2006. The research program of historical ecology. *Annual Review of Anthropology* 35(1): 75–98.

Barnett, E. 1861. *Testimony in U.S. v. George C. Yount*. Land case no. 32 ND [Caymus]. United States District Court, Northern District of California. Courtesy of The Bancroft Library, UC Berkeley.

Bartlett, J. R. 1854. *Personal narrative of explorations and incidents in Texas, New Mexico, California, Sonora, and Chihuahua, connected with the United States and Mexican Boundary Commission during the years 1850, '51, '52, and '53*. Reprint, Chicago: Rio Grande Press, 1965.

Bartlett, W. P. 1928. *More happenings in California: A series of sketches of the great California out-of-doors*. Boston: Christopher Publishing House.

Baye, P. R., P. M. Faber, and B. Grewell. 2000. "Tidal marsh plants of the San Francisco Estuary." In *Baylands ecosystem species and community profiles. Life histories and environmental requirements of key plants, fish and wildlife prepared by the San Francisco Bay Area Wetlands Ecosystem Goals Project*, 9–32. San Francisco and Oakland, CA: U.S. Environmental Protection Agency and S.F. Bay Regional Water Quality Control Board.

BCDC (San Francisco Bay Conservation and Development Commission). 2009. *Living with a rising bay: Vulnerability and adaptation in San Francisco Bay and on its shoreline*. Draft staff report, April 7, 2009. San Francisco: BCDC.

Beagle, J. R. 2010. *An anticipatory management plan for Carneros Creek, Napa, CA*. Unpublished MS thesis, University of California, Berkeley, CA.

Becker, G. S., I. J. Reining, D. A. Asbury, and A. Gunther. 2007. *San Francisco Estuary watersheds evaluation*. Oakland, CA: Center for Ecosystem Management and Restoration.

Beechie, T. J., M. Liermann, M. M. Pollock, S. Baker, and J. Davies. 2006. "Channel pattern and river-floodplain dynamics in forested mountain river systems." *Geomorphology* 78: 124–141.

Belden D. 1887. "The Santa Clara Valley." *Overland monthly* 9(54): 561–577.

Beller, E. E., M. N. Salomon, and R. M. Grossinger. 2010. *Historical vegetation and drainage patterns of western Santa Clara Valley: A technical memorandum describing landscape ecology in Lower Peninsula, West Valley, and Guadalupe Watershed management areas.* SFEI Contribution 622. Oakland, CA: San Francisco Estuary Institute.

Benner, P. A. and J. R. Sedell. 1997. "Upper Willamette River landscape: A historic perspective." In *River quality: Dynamics and restoration*, eds. A. Laenen and D.A. Dunnette. Boca Raton, FL: CRC Press.

Berner, M., B. Grummer, R. Leong, and M. Rippey. 2003. *Breeding birds of Napa County, California: An illustrated atlas of nesting birds*, ed. A. Smith. Vallejo, CA: Napa-Solano Audubon Society.

Bickford, E. L. 1928 (published 1929). "Notes from Napa Valley." *The Condor* 31(1): 35–36.

Bickford, E. L. 1931 (published 1932). "The wood ducks in Napa County, California." *The Condor* 34(2): 101.

Bock, C. E. and J. H. Bock. 2000. *The views from Bald Hill: Thirty years in Arizona grassland.* Berkeley, CA: University of California Press.

Bolliger, J., L. A. Schulte, S. N. Burrows, T. A. Sickley, and D. J. Mladenoff. 2004. "Assessing ecological restoration potentials of Wisconsin (U.S.A.) using historical landscape reconstructions." *Restoration Ecology* 12(1): 124–143.

Boughton D. A., P. B. Adams, E. Anderson, C. Fusaro, E. Keller, E. Kelley, L. Lentsch, J. Nielsen, K. Perry, H. Regan, J. Smith, C. Sift, L. Thompson, and F. Watson. 2006. *Steelhead of the South-Central/Southern California Coast: Population characterization for recovery planning*, report no. NOAA-TM-NMFS-SWFSC-394. U.S. Department of Commerce, National Oceanic and Atmospheric Administration, National Marine Fisheries Service, Southwest Fisheries Science Center.

Bousman, W. G. 2007. "Breeding avifaunal changes in the San Francisco Bay Area 1927–2005." *Western Birds* 38(2): 102–136.

Bowles, S. 1865. *Across the continent: A summer's journey to the Rocky Mountains, the Mormons, and the Pacific states, with Speaker Colfax.* Springfield: Samuel Bowles & Company.

Brace, C. L. 1869. *The new west: California in 1867–1868.* New York: G. P. Putnam & Sons.

Brewer, W. H. 1966. *Up and down California in 1860–1864: The journal of William H. Brewer.* New [3rd] edition, ed. F. P. Farquhar. Berkeley, CA: University of California Press, 1974.

Brierley, G. J., T. Cohen, K. Fryirs, and A. Brooks. 1999. "Post-European changes to the fluvial geomorphology of Bega catchment, Australia: Implications for river ecology." *Freshwater Biology* 41: 839–848.

Briggs, L. V. 1931. *California and the West, 1881 and later.* Boston: Wright & Potter Printing Company.

Brown, A. K. 2002. *Historical oak woodland detected through Armillaria mellea damage in fruit orchards.* U.S. Forest Service general technical report PSW-GTR-184: 651–661.

Brown, A. K. 2005. *Reconstructing early historical landscapes in the Northern Santa Clara Valley*, ed. R. K. Skowronek. Santa Clara, CA: Santa Clara University.

Brown, R. and F. W. Davis. 1991. "Historical mortality of valley oak (*Quercus lobata nee*) in the Santa Ynez Valley, Santa Barbara County, 1933–1989." In *Proceedings of oak woodlands and hardwood rangeland management: A research symposium*. U.S. Forest Service General Technical Report PSW-126, 202–207.

Bryan, E. N. 1932. *Report of Napa Valley investigation.* California Department of Public Works. Sacramento, CA: California State Printing Office.

Buckman, O. H. 1886. *Survey of the several tracts of land comprising the estate of John Tychson, near St. Helena.* Courtesy of Napa County Surveyor's Office.

Bucknall, M. E. 1917. "The days of long ago." *St. Helena Star, April 6, 1917.* Reprinted in *Gleanings*, vol. 4, no. 4. Napa, CA: Napa County Historical Society, 1993.

Byrns, R. W. 1922. *Descriptive report. Topographic sheet no. 4020, locality San Pablo, revision of Napa River*. Washington DC: U.S. Coast and Geodetic Survey.

California Senate. 1859. *Appendix to journals of the Senate of the tenth session of the Legislature of the state of California*. Sacramento, CA: John O'Meara, State Printer for California.

California Star. 1848a. "Inland navigation," February 12. From California Digital Newspaper Collection, http://cdnc.ucr.edu/about_us.html.

California Star. 1848b. "A trip across the bay," January 22. From California Digital Newspaper Collection, http://cdnc.ucr.edu/about_us.html.

California Star. 1848c. "A trip across the bay," January 29. From California Digital Newspaper Collection, http://cdnc.ucr.edu/about_us.html.

Camp, C. L. 1966. *George C. Yount and his chronicles of the west*. Denver, CO: Old West Publishing Company.

Carpenter, E. J. and S. W. Cosby. 1938. *Soil survey of the Napa area, California*. Series 1933, no. 13. U.S. Department of Agriculture, Bureau of Chemistry and Soils.

Caton, J. D. 1879. *Miscellanies*. Boston, MA: Houghton, Osgood and Company.

Cayan, D. R., K. Nicholas, M. Tyree, and M. D. Dettinger. 2011. *Climate and phenology in Napa Valley: A compilation and analysis of historical data*. Napa, CA: Napa Valley Vintners Association.

CDFG (California Department of Fish and Game). ca. 1979. *Napa River fisheries flow requirement study*. Sacramento, CA: CDFG, Bay Delta Region.

CDWR (California Department of Water Resources). 1995. *Historical ground water levels in Napa Valley*. Sacramento, CA: CDWR.

CLSI (California Land Stewardship Institute). 2011. *Napa river restoration: Oakville to Oak Knoll. Final concept plan*. Napa, CA: CLSI.

Cluer, B., M. Swanson, and J. McKeon. 2009. *Ecological opportunities for gravel pit reclamation on the Russian River*. Presented at Russian River Pit Symposium, Fountain Grove Inn, Santa Rosa, CA.

Clyman, J. 1960. *James Clyman, Frontiersman. 1792–1881. The adventures of a trapper and covered-wagon emigrant as told in his own reminiscences and diaries*, ed. C. L. Camp. Portland, OR: Champoeg Press.

Clyman, J. and R. J. Montgomery. ca. 1846. *Diary of Col. James Clyman*. Courtesy of The Bancroft Library, UC Berkeley.

CNDDB (California Natural Diversity Database), www.dfg.ca.gov/biogeodata/cnddb. California Department of Fish and Game, Wildlife and Habitat Data Analysis Branch, Habitat Conservation Division.

Collins, B. D. and D. R. Montgomery. 2001. "Importance of archival and process studies to characterizing pre-settlement riverine geomorphic processes and habitat in the Puget lowland." *Water Science and Application* 4: 227–243.

Collins, B. D. and D. R. Montgomery. 2002. "Forest development, wood jams, and restoration of floodplain rivers in the Puget lowland, Washington." *Restoration Ecology* 10(2): 237–247.

Collins, B. D., D. R. Montgomery, and A. D. Haas. 2002. "Historical changes in the distribution and functions of large wood in Puget lowland rivers." *Canadian Journal of Fisheries and Aquatic Sciences* 56: 66–76.

Collins, B. D., D. R. Montgomery, and A. J. Sheikh. 2003. "Reconstructing the historical riverine landscape of the Puget lowland." In *Restoration of Puget Sound rivers*, ed. D. R. Montgomery, 79–128. Seattle: Center for Water and Watershed Studies in association with University of Washington Press.

Collins, J. N., J. L. Grenier, J. Didonato, G. Geupel, T. Kucera, B. Lidicker, B. Rainey, and S. Rottenborn. 2007. *Ecological connections between baylands and uplands: Examples from Marin County*. SFEI Contribution 521. Oakland, CA: San Francisco Estuary Institute.

Collins, J. N., M. Sutula, E. D. Stein, M. Odaya, E. Zhang, and K. Larned. 2006. *Comparison of methods to map California riparian areas. Final report prepared for the California Riparian Habitat Joint Venture*. SFEI Contribution 522. Oakland and Costa Mesa, CA: San Francisco Estuary Institute and Southern California Coastal Water Research Project.

Cone, M. 1876. *Two years in California*. Chicago, IL: S. C. Griggs and Company.

Cook, S. F. 1956. *The Aboriginal population of the north coast of California*. University of California Anthropological Records, vol. 16, no. 3. Berkeley, CA.

Cooke, R. U. and R. W. Reeves. 1976. *Arroyos and environmental change in the American South-West*. Oxford: Clarendon Press.

Coombs, N. 1861. *Testimony in U.S. v. George C. Yount*. Land case no. 32 ND, 558–559. District Court of the United States, Northern District of California. Courtesy of The Bancroft Library, UC Berkeley.

Cooper, W. S. 1926. "The fundamentals of vegetational change." *Ecology* 7(4): 391–413.

Costello, L., B. W. Hagen, and K. S. Jones. 2011. *Oaks in the urban landscape: Selection, care, and preservation*. Publication 3518. Richmond, CA: University of California Agriculture and Natural Resources.

Cowardin L. M., V. Carter, F. C. Golet, and E. T. LaRoe. 1979. *Classification of wetlands and deepwater habitats of the United States*. Washington DC: Fish and Wildlife Service, Biological Services Program, U.S. Department of the Interior.

Cronise, T. F. 1868. *The natural wealth of California*. San Francisco: H. H. Bancroft & Company.

Cronon W. 1983. *Changes in the land. Indians, colonists and the ecology of New England*. New York: Hill and Wang.

Curtis, E. J. 1924. *A Wappo in the North American Indian: Being a series of volumes picturing and describing the Indians of the United States, and Alaska*, vol. 14, plate 490. Seattle, WA and Cambridge, MA: E. S. Curtis and Cambridge University Press.

Cuthrell, R., C. J. Striplen, and K. G. Lightfoot. 2009. "Exploring indigenous landscape management at Quiroste Valley, the archaeological approach." *News from Native California* 22(3): 26–29.

Daily Alta California. 1857. "A prospecting tourist," July 31. From California Digital Newspaper Collection, http://cdnc.ucr.edu/about_us.html.

Daily Alta California. 1860. "Notes on Napa Valley: Wheat and granaries," February 21. From California Digital Newspaper Collection, http://cdnc.edu/about_us.html.

Daily Alta California. 1864. "On the way to Calistoga," August 18. From California Digital Newspaper Collection, http://cdnc.edu/about_us.html.

Darms, H. A. 1908. *Napa City & County Portfolio & Directory*. Napa, CA: H. A. Darms & Gordon Eby.

Dettinger, M. D., D. R. Cayan, H. F. Diaz, and D. M. Meko. 1998. "North-south precipitation patterns in western North America on interannual-to-decadal timescales." *Journal of Climate* 11(12): 3095–3111.

Dewoody, T. J. 1860. *Survey of the Caymus Rancho and adjoining portions of Rancho Napa and Rancho Yajome: Napa Co., Calif*. Land case no. 32 ND, map E-70. District Court of the United States, Northern District of California. Courtesy of The Bancroft Library, UC Berkeley.

Dewoody, T. J. 1866a. *Calistoga hot springs, and adjoining land. Napa County, Cal. Samuel Brannan Esq., Proprietor*. Courtesy of Napa County Surveyor's Office.

Dewoody, T. J. 1866b. *Field notes of the final survey of the Rancho Carne Humana, the heirs of Edward Bale, Confirmee*. Land case no. 47 ND. General Land Office, U.S. Department of the Interior, Bureau of Land Management Rectangular Survey, California, vol. G5. Courtesy of Bureau of Land Management, Sacramento, CA.

Dewoody, T. J. 1867. *Field notes of the survey of exterior boundaries of Township 8 North, Range 6 West, Mt. Diablo Meridian*. General Land Office, U.S. Department of the Interior, Bureau of Land Management Rectangular Survey, California, vol. R179. Courtesy of Bureau of Land Management, Sacramento, CA.

Dewoody, T. J. 1870. *Map of the Subdivision of the Caymus Grant in Napa County, California. Yount Estate*. Courtesy of The Bancroft Library, UC Berkeley.

Dewoody, T. J. 1873. *Mill Tract portion of Carne Humana Rancho, Napa County, California*. Napa, CA: Haas Bros. Stationers and Lithographers.

Dewoody, T. J. 1879. *Plat of Part of the Rancho Entre Napa finally confirmed to Joseph Green*. Land case no. 172 ND. Courtesy of California State Lands Commission.

D'Hémécourt, E. 1855. *Survey of property of John Truebody*, survey no. 47. In *Brands*, vol. A, 97. Courtesy of Napa County Surveyor's Office.

Dillon, R. H. 2001. *Napa Valley's natives*. Fairfield, CA: James Stevenson Publisher.

Dillon, R. H. 2004. *Napa Valley heyday*. San Francisco: The Book Club of California.

Dillon, R. H. n.d. *Building eden*. Unpublished manuscript. *Courtesy of Napa County Historical Society*.

Dilts, T. E. Forthcoming. "Using historical General Land Office survey notes to quantify the effects of irrigated agriculture on land cover change in an arid lands watershed." *Annals of the Association of American Geographers*.

Driver, H. 1936. "Wappo ethnography." *University of California Publications in American Archaeology and Ethnography* 36: 179–220.

Duhig, S. M. 1990. *Huichica: Recollections of Stewart M. Duhig*, ed. V. Duhig. Napa, CA: Duhig Publications.

Dunne, T. and L. B. Leopold. 1978. *Water in environmental planning*. New York: W. H. Freeman and Company.

Egan, D. and E. A. Howell, eds. 2005. *The historical ecology handbook: A restorationist's guide to reference ecosystems*. Washington, DC: Island Press.

Erlandson, J. M. and T. L. Jones. 2002. *Catalysts to complexity: Late holocene societies of the California coast*, ed. J. E. Arnold. Los Angeles: Cotsen Institute of Archaeology, University of California.

Evans, A. S. 1873. *A la California: Sketches of life in the Golden State*. San Francisco: A. L. Bancroft & Company.

Faber, P. M. 2005. *California's wild gardens: A guide to favorite botanical sites*. Berkeley, CA: University of California Press.

Faye, R. E. 1973. *Ground-water hydrology of northern Napa Valley, California*. Menlo Park, CA: U.S. Geological Survey.

Fels T. W. 2005. *Documenting eden*. San Francisco, CA: The Society of California Pioneers.

FGDC (Federal Geographic Data Committee). 1997. *Vegetation classification standard*. Reston, VA: U.S. Geological Survey.

Fisher, C. K. 1959. *Napa River stream survey by Fisher 1-20-59; creel census WPB 1954–55 winter steelhead season*. California Department of Fish and Game.

Foster, D. R. 2000. "From boblinks to bears: Interjecting geographic history into ecological studies, environmental interpretation, and conservation planning." *Journal of Biogeography* 27: 27–30.

Foster, D. R. 2002. "Insights from historical geography to ecology and conservation: Lessons from the New England landscape." *Journal of Biogeography* 29: 1269–1275.

Foster, D. R. and G. Motzkin. 2003. "Interpreting and conserving the openland habitats of coastal New England: Insights from landscape history." *Forest Ecology and Management* 185: 127–150.

Fryirs, K. 2010. "Post-European settlement disturbance response of rivers in Bega catchment, South Coast, NSW, Australia." In *Key concepts in geomorphology*, eds. P. Bierman and D. R. Montgomery. Burlington, VT: W. H. Freeman.

Gardner, D. 1977. *Suscol in Napa County: An historic report 1835–1977*. Courtesy of Sonoma County Library.

Geiger, J. 2004. "The large tree argument: The case for large trees vs. small trees." *Western Arborist*: 14–15.

Gilbert, G. K. 1917. *Hydraulic-mining debris in the Sierra Nevada*. U.S. Geological Survey professional paper no. 105. Washington, DC: U.S. Government Printing Office.

Gillette, R. 1972. "Stream channelization: Conflict between ditchers, conservationists." *Science* 176(4037): 890–894.

Goals Project. 1999. *Baylands ecosystem habitat goals. A report of habitat recommendations prepared by the San Francisco Bay Area Wetlands Ecosystem Goals Project*. San Francisco and Oakland, CA: U.S. Environmental Protection Agency and S.F. Bay Regional Water Quality Control Board.

Graf, W. L. 2000. "Locational probability for a dammed, urbanizing stream: Salt River, Arizona, USA." *Environmental Management* 25(3): 321.

Gray, N. 1853. *Field notes of the survey of a portion of the standard parallel and township lines from no. 6 to 8 and range 1 to 6 west of the meridian*, General Land Office, U.S. Department of the Interior, Bureau of Land Management Rectangular Survey, California, vol. R254. Courtesy of Bureau of Land Management, Sacramento, CA.

Graymer, R. W., B. C. Moring, G. J. Saucedo, C. M. Wentworth, E. E. Brabb, and K. L. Knudsen. 2006. *Geologic map of the San Francisco Bay region, U.S. Geological Survey scientific investigations map 2918*. Menlo Park, CA: U.S. Geological Survey.

Grayson, A. J. 1986. *Birds of the Pacific slope: One hundred fifty-six bird portraits painted in California and Mexico 1853–1869*. San Francisco: Arion Press.

Gregory, T. 1912. *History of Solano and Napa counties, California*. Los Angeles, CA: Historic Record Co.

Grinnell, J. 1936. *Field notes: Nov 26–29, Napa Co., Calif*. Museum of Vertebrate Zoology Archival Field Notes, http://bscit.berkeley.edu/mvz/volumes.html, 2680–2687.

Grinnell, J. H. C. Bryant, and T. I. Storer. 1918. *The game birds of California*. Berkeley, CA: University of California Press.

Grinnell, J., J. S. Dixon, and J. M. Linsdale. 1937. *Fur-bearing mammals of California: Their natural history, systematic status, and relations to man*. Berkeley, CA: University of California Press.

Grossinger, R. M. 1995. *Historical evidence of freshwater effects on the plan form of tidal marshlands in the Golden Gate Estuary*. MS thesis, University of California, Santa Cruz.

Grossinger, R. M. 2005. "Documenting local landscape change: The San Francisco Bay area historical ecology project." In *The historical ecology handbook: A restorationist's guide to reference ecosystems*, eds. D. Egan and E. A. Howell, 425–442. Washington, DC: Island Press.

Grossinger, R. M., R. A. Askevold, C. J. Striplen, E. Brewster, S. Pearce, K. N. Larned, L. J. McKee, and J. N. Collins. 2006. *Coyote Creek watershed historical ecology study: Historical condition, landscape change, and restoration potential in the eastern Santa Clara Valley, California*. SFEI Contribution 426. Oakland, CA: San Francisco Estuary Institute.

Grossinger, R. M. and E. E. Beller. 2007. *Landscape history of the Trancas, technical memorandum to Design, Community, and Environment*. Oakland, CA: San Francisco Estuary Institute.

Grossinger, R. M. and E. E. Beller. 2011. "Oak landscapes in the recent past." In *Oaks in the urban landscape*, 222–223. Richmond, CA: University of California Agriculture and Natural Resources.

Grossinger, R. M., E. E. Beller, M. N. Salomon, A. A. Whipple, R. A. Askevold, C. J. Striplen, E. Brewster, and R. A. Leidy. 2008. *South Santa Clara historical ecology study, including Soap Lake, the Upper Pajaro River, and Llagas, Uvas-Carnadero, and Pacheco Creeks. Prepared for the Santa Clara Valley Water District and the Nature Conservancy*. SFEI Contribution 558. Oakland, CA: San Francisco Estuary Institute.

Grossinger, R. M., C. J. Striplen, R. A. Askevold, E. Brewster, and E. E. Beller. 2007. "Historical landscape ecology of an urbanized California valley: Wetlands and woodlands in the Santa Clara Valley." *Landscape Ecology* 22: 103–120.

Grossinger, R. M., C. J. Striplen, E. Brewster, and L. McKee. 2004a. *Ecological, geomorphic, and land use history of Carneros Creek watershed: A component of the watershed management plan for the Carneros Creek watershed, Napa County, California. A technical report of the Regional Watershed Program*. SFEI Contribution 70. Oakland, CA: San Francisco Estuary Institute.

Grossinger, R. M., C. J. Striplen, E. Brewster, and L. McKee. 2004b. *Ecological, geomorphic, and land use history of Sulphur Creek watershed: A component of the watershed management plan for the Sulphur Creek watershed, Napa County, California. A technical report of the Regional Watershed Program*. SFEI Contribution 69. Oakland, CA: San Francisco Estuary Institute.

Gunn, H. L. and M. Hunt. 1926. *History of Solano County and Napa County*. Chicago: S. J. Clarke.

Hall, H. M. 1923. *Record for Parry's tarplant (Centromadia parryi* subsp. *parryi)*, accession no. UC406892. From *Consortium of California Herbaria*, http://ucjeps.berkeley.edu/consortium.

Hamilton, J. G. 1997. "Changing perceptions of pre-European grasslands in California." *Madroño* 44(4): 311–333.

Harley, J. B. 1990. "Text and contexts in the interpretation of early maps." In S*ea charts to satellite images: Interpreting North American history through maps*, ed. D. Buisseret, 3–15. Chicago: University of Chicago Press.

Harwood K.and A. G. Brown. 1993. "Fluvial processes in a forested anastomosing river: Flood partitioning and changing flow patterns." *Earth Surface Processes and Landforms* 18(8):741-748.

Heberger, M., H. Cooley, P. Herrera, P. H. Gleick, and E. Moore. 2009. *The impacts of sea-level rise on the California coast*. Oakland, CA: The Pacific Institute.

Helley, E. J. and K. R. LaJoie. 1979. *Flatland deposits of the San Francisco Bay region, California: Their geology and engineering properties and their importance to comprehensive planning*. Washington, DC: Government Printing Office.

Henning, J., R. E. Gresswell, and I. A. Fleming. 2006. "Juvenile salmonid use of freshwater emergent wetlands in the floodplain and its implications for conservation management." *North American Journal of Fisheries Management* 26: 367–376.

Hittell, J. S. 1863. *The resources of California*. San Francisco: A. Roman & Company.

Holland R. F. 1986. *Preliminary descriptions of the terrestrial natural communities of California*. Unpublished report. Sacramento, CA: California Department of Fish and Game, Natural Heritage Division.

Holmes, L. C. and J. W. Nelson. 1917. U.S. Department of Agriculture: Bureau of Soils. *Reconnaissance soil survey of the San Francisco Bay region, California*. Washington, DC: Government Printing Office.

Holstein, G. 2000. "Plant communities ecotonal to the Baylands." In *Baylands ecosystem species and community profiles. Life histories and environmental requirements of key plants, fish and wildlife. Prepared by the San Francisco Bay Area Wetlands Ecosystem Goals Project*, ed. P. Olofson, 49–68.

San Francisco and Oakland, CA: U.S. Environmental Protection Agency and San Francisco Bay Regional Water Quality Control Board.

Holstein, G. 2001. "Pre-agricultural grassland in Central California". *Madroño* 48(4): 253–264.

Houghton, J. F. 1862. *Annual Report of the Surveyor-General of California for the year 1862.* Sacramento, CA: Surveyor General's Office.

Hulse, D., S. Gregory, and J. P. Baker. 2002. *Willamette River Basin planning atlas: Trajectories of environmental and ecological change.* Corvallis: Oregon State University Press.

Husmann G. C. 1912. "Viticulture of Napa County." In *A history of Solano and Napa counties, California: With biographical sketches of the leading men and women of the counties, who have been identified with its growth and development from the early days to the present time.* Los Angeles, CA: Historic Record Company.

Ingols, R. M. 2006. *Back of the beyond: An anthology of memories,* ed. R. Ingols. Napa, CA: California Native Plant Society, California Chapter.

Jeffres, C. A., J. J. Opperman, and P. B. Moyle. 2008. "Ephemeral flooplain habitats provide best growth conditions for juvenile Chinook salmon in a California river." *Environmental Biology of Fishes* 83: 449–458.

Jepson, W. L. 1892. *Field book of Willis L. Jepson,* vol. 48. Courtesy of Jepson Herbarium, University of California, Berkeley, http://ucjeps.berkeley.edu/images/fieldbooks/jepson_fieldbooks.html.

Jepson, W. L. 1893. *Records for rusty popcorn flower (*Plagiobothrys nothofulvus*), collection nos. 21162, 21156, and 21154.* Courtesy of Consortium of California Herbaria, http://ucjeps.berkeley.edu/consortium.

Jepson, W. L. 1897. *Record for rusty popcorn flower (*Plagiobothrys nothofulvus*), collection no. 21158.* Courtesy of Consortium of California Herbaria, http://ucjeps.berkeley.edu/consortium.

Jepson, W. L. 1900. *Field book of Willis L. Jepson,* vol. 3. Courtesy of Jepson Herbarium, University of California, Berkeley, http://ucjeps.berkeley.edu/images/fieldbooks/jepson_fieldbooks.html.

Jepson, W. L. 1903. *Field book of Willis L. Jepson,* vol. 11. Courtesy of Jepson Herbarium, University of California, Berkeley, http://ucjeps.berkeley.edu/images/fieldbooks/jepson_fieldbooks.html.

Jepson, W. L. 1909. *The trees of California.* San Francisco: Cunningham, Curtis & Welch.

Jepson, W. L. 1910. "The silva of California." In *Memoirs of the University of California,* vol. 2. Berkeley, CA: University of California Press.

Jepson, W. L. 1912. "The trees, shrubs, and flowers of Napa Valley." In *History of Solano and Napa Counties,* ed. T. Gregory. Los Angeles: Historic Record Company.

Jepson, W. L. 1918. *Field book of Willis L. Jepson,* vol. 18. Courtesy of Jepson Herbarium, University of California, Berkeley, http://ucjeps.berkeley.edu/images/fieldbooks/jepson_fieldbooks.html.

Jepson, W. L. 1922. *Field book of Willis L. Jepson,* vol. 39. Courtesy of Jepson Herbarium, University of California, Berkeley, http://ucjeps.berkeley.edu/images/fieldbooks/jepson_fieldbooks.html.

Jepson, W. L. 1934. *Field book of Willis L. Jepson,* vol. 54. Courtesy of Jepson Herbarium, University of California, Berkeley, http://ucjeps.berkeley.edu/images/fieldbooks/jepson_fieldbooks.html.

Jepson, W. L. 1936a. *A flora of California.* Berkeley, CA: Jepson Herbarium and Library, University of California.

Jepson W. L. 1936b. *Field book of Willis L. Jepson,* vol. 56. Courtesy of Jepson Herbarium, University of California, Berkeley, http://ucjeps.berkeley.edu/images/fieldbooks/jepson_fieldbooks.html.

Johnston, W. E. and A. F. McCalla. 2004. *Whither California agriculture: Up, down, or out? Some thoughts about the future.* Special report 04-1. Davis, CA: Giannini Foundation of Agricultural Economics, University of California.

Kerr, D. 1858. *Napa Creek and Napa City California, plane table sheet XXXII, register no. T777.* Washington, DC: U.S. Coast Survey.

Ketteringham. 1961. *Settlement geography of the Napa Valley.* MS thesis, Stanford University, Stanford, CA.

King, M. G. and T. W. Morgan. 1881. *Map of the central portion of the Napa Valley and the town St. Helena.* Courtesy of California State Library, California History Center.

Knight, W. 1980. "The story of Browns Island." *Four Seasons* 6(1): 8.

Knighton, D. 1998. *Fluvial forms and processes: A new perspective.* New York: Hodder-Arnold.

Knowles, N. 2010. "Potential inundation due to rising sea levels in the San Francisco Bay region." *San Francisco Estuary and Watershed Science* 8(1).

Koehler, J. and P. Blank. 2010. *Napa River steelhead and salmon smolt monitoring program annual report—year 2.* Napa, CA: Napa County Resource Conservation District.

Kondolf, M. G. 2006. "River restoration and meanders." *Ecology and Society* 11(2): 42–60.

Kondolf, M. G., H. Piegay, and N. Landon. 2007. "Changes in the riparian zone of the lower Eygues River, France, since 1830." *Landscape Ecology* 22: 367–384.

Kondolf G. M., M. W. Smeltzer, and S. F. Railsback. 2001. "Design and performance of a channel reconstruction project in a coastal California gravel-bed stream." *Environmental Management* 28(6):761-776.

Kunkel, F. and J. E. Upson. 1960. U.S. Department of the Interior Geological Survey and California Department of Water Resources. *Geology and ground water in Napa and Sonoma Valleys, Napa and Sonoma Counties, California,* 252. Washington, DC: US Government Printing Office.

Lambert, G. and J. H. Kashiwagi. 1978. *Soil Survey of Napa County, California.* Washington, DC: U.S. Department of Agriculture.

Lapham, M. H. 1949. *Crisscross trails: Narrative of a soil surveyor.* Berkeley, CA: W. E. Berg.

Laymon, S. A. 1998. "Yellow-billed cuckoo (*Coccycus americanus*)." In *The riparian bird conservation plan: A strategy for reversing the decline of riparian-associated birds in California.* Petaluma, CA: Point Reyes Bird Observatory.

Laymon, S. A. and M. D. Halterman. 1989. *A proposed habitat management plan for Yellow-billed Cuckoos in California.* Presented at California Riparian Systems Conference, U.S. Department of Agriculture, Davis, CA.

Leach, F. A. 1917. *Recollections of a newspaperman.* San Francisco: Samuel Levinson.

Lebassi B., J. Gonzalez, D. Fabris, E. Maurer, N. Miller, C. Milesi, P. Switzer, and R. Bornstein. 2009. "Observed 1970-2005 cooling of summer daytime temperatures in coastal California." *Journal of Climate* 22.

Leidy, R. A. 2007. *Ecology, assemblage structure, distribution, and status of fishes in streams tributary to the San Francisco Estuary, California.* SFEI Contribution 530. Oakland, CA: San Francisco Estuary Institute.

Leidy, R. A., G. S. Becker, and B. N. Harvey. 2005a. *Historical distribution and current status of steelhead/rainbow trout* (Oncorhynchus mykiss) *in streams of the San Francisco Estuary, California.* Oakland, CA: Center for Ecosystem Management and Restoration.

Leidy, R. A., G. S. Becker, and B. N. Harvey BN. 2005b. "Historical status of Coho salmon in streams of the urbanized San Francisco Estuary, California." *California Fish and Game* 91(4): 35.

Leidy R. A., K. Cervantes-Yoshida, and S. Carlson. 2011. "Persistence of native fishes in small streams of the urbanized San Francisco Estuary, California: acknowledging the role of urban streams in native fish conservation." *Aquatic Conservation: Marine and Freshwater Ecosystems* 21:472-483.

Leopold, L. B. 1994. *A view of the river.* Cambridge, MA: Harvard University Press.

Lightfoot, K. G. and O. Parrish. 2009. *California Indians and their environment: An introduction.* Berkeley, CA: University of California Press.

Lightfoot, K. G., C. Striplen, and R. Cuthrell. 2009. "The Collaborative Research Program at Quiroste Valley." *News from Native California* 22(2): 30–33.

Loring. 1853. *Map of part of Napa County known as Nicolas Higuera's Tract [part of Rancho Entre Napa, Calif.]*. Land case map E-394. U.S. District Court, Northern District. Courtesy of The Bancroft Library, UC Berkeley.

Malamud-Roam, F. P., M. Dettinger, B. Ingram, M. Hughes, and J. Florsheim. 2007. "Holocene climates and connections between the San Francisco Bay estuary and its watershed: A review." *San Francisco Estuary and Watershed Science* 5(1): Article 3.

Malamud-Roam, F. P., B. L. Ingram, M. Hughes, and J. L. Florsheim. 2006. "Holocene paleoclimate records from a large California estuarine system and its watershed region: Linking watershed climate and bay conditions." *Quaternary Science Reviews* 25(13–14): 1570–1598.

Maloney, A. B. 1943. "John Work of the Hudson's Bay Company: Leader of the California brigade of 1832–33." *California Historical Society Quarterly* 22(2): 97–109.

Manies, K. L. and D. J. Mladenoff. 2000. "Testing methods to produce landscape-scale presettlement vegetation maps from the U.S. public land survey records." *Landscape Ecology* 15: 741–754.

Manning, A. D., J. Fischer, and D. B. Lindenmayer. 2006. "Scattered trees are keystone structures— implications for conservation." *Biological Conservation* 132: 311–321.

Marsh, G. P. 1869. *Man and nature; or, Physical geography as modified by human action*. New York: C. Scribner & Co.

Mayer, K. E. and W. F. Laudenslayer Jr. 1988. *A guide to wildlife habitats of California*. California Department of Fish and Game.

McClellan R. G. 1872. *The Golden State: A history of the region west of the mountains; Embracing California, Oregon, Nevada, Utah, Arizona, Idaho, Washington Territory, British Columbia and Alaska from the earliest period to the present time*. San Francisco, CA: A. Roman & Co.

McGlashan, H. D. and H. J. Dean. 1913. *Water resources of California, Part III. Stream measurements in the Great Basin and Pacific Coast river basins*. U.S. Geological Survey, water-supply paper 300. Washington, DC: Government Printing Office.

McPherson, G. 2007. "Urban tree planting and greenhouse gas reductions." *Arborist News*: 32–34.

Mendell, G. 1885. "Appendix PP: The improvement of the harbors of Oakland and Wilmington; of Petaluma Creek, and of the harbor of Redwood, California." *Annual report of the Chief of Engineers, United States Army to the Secretary of War*, 1985. Washington, DC: Government Printing Office.

Menefee, C. A. 1873. *Historical and descriptive sketch book of Napa, Sonoma, Lake and Mendocino: comprising sketches of their topography, productions, history, scenery, and peculiar attractions*. Reprinted, Fairfield, CA: James Stevenson, Publisher, 1965.

Micheli, E. R. and J. W. Kirchner. 2002. "Effects of wet meadow riparian vegetation on streambank erosion 2. Measurements of vegetated bank strength and consequences for failure mechanics." *Earth Surface Processes and Landforms* 27: 687–697.

Michener, Bioletti. 1891. *Record for showy indian clover (Trifolium amoenum)*, accession no. UC85329. From *Consortium of California Herbaria*, http://ucjeps.berkeley.edu/consortium.

Milliken, R. A. 1978. "Ethno-history of the lower Napa Valley." In *Report of archaeological excavations at the River Glen site (CA-NAP-261), Napa County, California. Prepared for the U.S. Army Corps of Engineers*. San Francisco: Archaeological Consulting and Research Services, Inc.

Milliken, R. A. 1995. *A time of little choice: The disintegration of tribal culture in the San Francisco Bay area, 1769–1810*. Menlo Park, CA: Ballena Press.

Milliken, R. A. 2009. *Ethnohistory and ethnogeography of the Coast Miwok and their neighbors, 1783–1840. A technical report prepared for the National Park Service, Golden Gate National Resources*

Area, Cultural Resources and Museum Management Division. Oakland, CA: Archaeological/Historical Consultants.

Minnich, R. A. 2008. *California's fading wildflowers: Lost legacy and biological invasions.* Berkeley, CA: University of California Press.

Mitsch, W. J. and J. G. Gosselink. 2000. "The value of wetlands: Importance of scale and landscape setting." *Ecological Economics* 35(1).

Montgomery, D. R. 2008. "Dreams of natural streams." *Science* 319(5861): 291–292.

Morgan, T. W. 1871. *Map of Calistoga or little geysers and the hot sulphur springs, Napa County, California.* Courtesy of California State Library History Room.

Moss, M. L. and J. M. Erlandson. 1995. "Reflections on North American Pacific coast prehistory." *Journal of World Prehistory* 9(1).

Mount, J. F. 1995. *California rivers and streams.* Berkeley, CA: University of California Press.

Mow, F. 1868. *Map of the Napa Valley Railroad Homestead Association, Calistoga, Napa County, Cal.*

Moyle, P. B., P. K. Crain, and K. Whitener. 2007. "Patterns in the use of a restored California floodplain by native and alien fishes." *San Francisco Estuary and Watershed Science* 5(3): Article 1.

Naiman R. J., C. A. Johnston, and J. C. Kelly. 1989. "Alteration of North American streams by beaver: The structure and dynamics of streams are changing as beaver recolonize their historic habitat." *BioScience* 38(11):753-762.

Napa County. 1921. *Plan for proposed reinforced concrete highway bridge over Napa river between St. Helena and the St. Helena sanitarium.* Sheet 1 of 4. Napa, CA: Napa County Surveyor's Office.

Napolitano, M. N., S. Potter, and D. Whyte. 2009. *Napa River watershed sediment TMDL and habitat enhancement plan.* Oakland, CA: San Francisco Bay Area Water Board.

NCALM (National Center for Airborn Laser Mapping). 2003. LiDAR metadata. http://opentopo.sdsc.edu/gridsphere/gridsphere?gs_action=lidarDataset&cid=geonlidarframeportlet&opentopoID=OTLAS.052010.26910.1.

NCFCWCD (Napa County Flood Control and Water Conservation District). 2005. *2050 Napa Valley Water Resources Study.* Napa, CA: NCFCWCD.

NCMAD (Napa County Mosquito Abatement District). ca.1934. *Project report—general, mosquito control, California, Buckly station—southwest of Napa.* Project SLF-101.

Nolte, G. S. 1959. *Watershed work plan for Napa River watershed.* Palo Alto, CA: George S. Nolte, Consulting Civil Engineers, prepared for the Napa County Flood Control and Water Conservation District and Napa County Soil Conservation District.

Noonan J., *Napa Valley Register*. 2011. "Census: Upvalley populations on decline." April 2. http://napavalleyregister.com. Accessed August 3, 2011.

Nordhoff, C. 1873. *California: For health, pleasure, and residence. A book for travellers and settlers.* New York: Harper & Brothers.

Olmstead, A. L. and P. W. Rhode. 2003. "The evolution of California agriculture, 1850–2000." In *California agriculture: Dimensions and issues*, ed. J. Siebert. Berkeley, CA: Giannini Foundation of Agricultural Economics, http://giannini.ucop.edu/CalAgbook.htm.

Pacific Rural Press. 1871. July 8. From California Digital Newspaper Collection. http://cdnc.ucr.edu/cdnc.

Pacific Rural Press. 1885. "Agricultural engineer: Reclamation of marsh lands in California." May 30. From California Digital Newspaper Collection. http://cdnc.ucr.edu/cdnc.

Parsons, M. E. 1909. *The wild flowers of California: Their names, haunts, and habits.* San Francisco: Cunningham, Curtiss & Welch.

Pavlik, B. M., P. C. Muick, S. G. Johnson, and M. Popper. 1991. *Oaks of California.* Los Olivos, CA: Cachuma Press, Inc.

Pearce, S. A. and R. M. Grossinger. 2004. "Relative effects of fluvial processes and historical land use on channel morphology in three sub-basins, Napa River basin, California, USA." In *Sediment transfer through the fluvial system*, eds. V. Golosov, V. Belyaev, and D. E. Walling. Wallingford, UK: International Association of Hydrological Sciences (IAHS).

Peattie, D. C. 1953. *A natural history of western trees*. New York: Bonanza Books.

Petts, G., H. Moller, and A. Roux, eds. 1989. *Historical change of large alluvial rivers: Western Europe*. Chichester, UK: John Wiley & Sons.

Pierce, W. 1866. *Plat of the change in the Napa City and Clear Lake Road from Younts Mill to J. M. Mayfield's corner*. Courtesy of Napa County Surveyor's Office.

Poland J. F. and R. L. Ireland. 1988. *Land subsidence in the Santa Clara Valley, California, as of 1982*. Professional Paper 497-F. Denver, CO: U.S. Geological Survey.

Pollock, M. M., T. J. Beechie, and C. E. Jordan. 2007. "Geomorphic changes upstream of beaver dams in Bridge Creek, an incised stream channel in the interior Columbia River basin, eastern Oregon." *Earth Surface Processes and Landforms* 32: 1174–1185.

Pollock, M. M., G. R. Pess, T. J. Beechie, and D. R. Montgomery. 2004. "The importance of beaver ponds to Coho salmon production in the Stillaguamish River Basin, Washington, USA." *North American Journal of Fisheries Management* 24: 749–760.

Rackham, O. 1994. *The illustrated history of the countryside*. London: Seven Dials.

Rees T. H. 1914. *Napa River, Cal. Letter from the secretary of water, transmitting, with a letter from the Chief of Engineers, reports on a preliminary examination and survey of Napa River, Cal., with a view to making the necessary cut-offs, consideration being given to any tender of cooperation on the part of local interests*. House of Representatives, document no. 795. From Water Resources Center Archives, UC Riverside and CSU San Bernardino. http://library.ucr.edu/wrca.

Revere, J. W. 1849. *A tour of duty in California*. New York: C. S. Francis & Co.

Rodgers, A. F. 1856. *San Francisco Bay, California, plane table sheet XXII, register no. 563*. Washington, DC: U.S. Coast Survey.

Rosgen, D. L. 1994. "A classification of natural rivers." *Catena* 22: 169–199.

Rosgen, D. L. 2007. "Rosgen geomorphic channel design." In *Part 654 Stream Restoration Design National Engineering Handbook*, 75. Washington, DC: U.S. Department of Agriculture, Natural Resources Conservation Service.

Ruygt, J. A. 1982a. *Record for alkali milkvetch (*Astragalus tener *var.* tener*), collection no. 1155*. From *Consortium of California Herbaria*, http://ucjeps.berkeley.edu/consortium.

Ruygt, J. A. 1982b. *Records for sack clover (*Trifolium depauperatum*), collection nos. 1215, 1173, and 1182*. From *Consortium of California Herbaria*, http://ucjeps.berkeley.edu/consortium.

Ruygt, J. A. 1991. *Record for San Joaquin spearscale (*Atriplex joaquiniana*), collection nos. 2839 and 2763*. From *Consortium of California Herbaria*, http://ucjeps.berkeley.edu/consortium/.

Ruygt, J. A. 2001. *Napa County vernal pools*. Unpublished manuscript.

Sacramento Daily Union. 1860. July 2. From California Digital Newspaper Collection, http://cdnc.ucr.edu/about_us.html.

Sacramento Daily Union. 1891. "Napa River improvements," March 25. From California Digital Newspaper Collection, http://cdnc.ucr.edu/about_us.html.

Saldi-Caromile, K., K. Bates, P. Skidmore, J. Barenti, and D. Pineo. 2004. *Stream habitat restoration guidelines: Final draft. Prepared for Washington State Aquatic Habitat Guidelines Program*. Olympia, WA: Washington Departments of Fish and Wildlife and Ecology and the U.S. Fish and Wildlife Service. http://wdfw.wa.gov/publications/pub.php?id=00043.

San Francisco Call. 1895a. "A beautiful valley," March 26. From California Digital Newspaper Collection, http://cdnc.ucr.edu/about_us.html.

San Francisco Call. 1895b. "Napa River examined, the Congressional Party examines its bends and bars," October 2. From California Digital Newspaper Collection, http://cdnc.ucr.edu/about_us.html.

San Francisco Call and Post. 1921. September 17.

Sanborn Map Company. 1910. *City of Napa, Napa Co., California.* New York: Sanborn Map Company.

Sanborn Perris Map Company. 1886. *City of Napa, Napa Co., California.* New York: Sanborn Perris Map Company.

Sanborn Perris Map Company. 1891. *City of Napa, Napa Co., California.* New York: Sanborn Perris Map Company.

Sanderson, E. W. 2009. *Mannahatta: A natural history of New York City.* New York: Harry N. Abrams, Inc.

Sapozhnikov, V. B. and E. Foufoula-Georgiou. 1999. "Horizontal and vertical self-organization of braided rivers." *Water Resources Research* 35(3): 843–851.

Sauer, C. 1930. "Historical geography and the Western frontier." In *Land and life: A selection of writings from Carl Ortwin Sauer*, ed. J. Leighly. Berkeley, CA: University of California Press, 1969.

Sauer, C. 1939. *A first book in geography: Man in nature, America before the days of the white men.* New York: Charles Scribner's Sons.

Sawyer, J. O., T. Keeler-Wolf, and J. Evens. 2009. *A manual of California vegetation.* Sacramento: California Native Plant Society.

Schiffman, P. M. 2007. "Species composition at the time of first European settlement." In *California grasslands: Ecology and management*, eds. M. R. Stromberg, J. D. Corbin, and C. M. D'Antonio. Berkeley, CA: University of California Press.

Schumm, S. A., M. D. Harvey and C. C. Watson. 1984. *Incised channel: Morphology, dynamics and control.* Littleton, CO: Water Resources Publications.

Scott, M. L., J. M. Friedman, and G. T. Auble. 1996. "Fluvial process and the establishment of bottomland trees." *Geomorphology* 14(4): 327–339.

Sear D. A., C. E. Millington, D. R. Kitts, and R. Jeffries. 2010. "Logjam controls on channels: Floodplain interactions in wooded catchments and their role in the formation of multi-channel patterns." *Geomorphology* 116(2010):305-319.

Sedell, J. R. and J. L. Froggatt. 1984. "Importance of streamside forests to large rivers: The isolation of the Willamette River, Oregon, U.S.A., from its floodplain by snagging and streamside forest removal." *Verh. Internat. Verein. Limnol.* 22: 1828–1834.

Sedell, J. R. and K. J. Luchessa. 1982. "Using the historical record as an aid to salmonid habitat enhancement." In *Acquisition and utilization of aquatic habitat inventory information*, ed. N. Armantrout. Bethesda, MD: America Fisheries Society.

Sedell, J. R., G. H. Reeves, R. F. Hauer, J. A. Stanford, and C. P. Hawkins. 1990. "Role of refugia in recovery from disturbances: Modern fragmented and disconnected river systems." *Environmental Management* 14: 711–724.

SFEI (San Francisco Estuary Institute). 1998. *Bay Area EcoAtlas: Geographic information system of wetland habitats past and present.* www.sfei.org/ecoatlas.

SFEI. 2011a. *The changing Napa River Valley: Past and present characteristics with implications for future management.* SFEI technical publication 615. Oakland, CA: SFEI.

SFEI. 2011b. *Bay Area aquatic resources inventory.* www.californiawetlands.net/tracker/ba/map.

Shalowitz, A. L. 1964. *Shore and sea boundaries, with special reference to the interpretation and use of Coast and Geodetic Survey data, United States*, ed. U.S. Department of Commerce and Coast and Geodetic Survey. Washington, DC: Government Printing Office.

Shiffman, P. M. 2007. "Species composition at the time of first European settlement." In *California grasslands: Ecology and management*, ed. M. Stromberg. Berkeley, CA: University of California Press.

Shortridge, C. M. 1896. *Santa Clara County and its resources: Historical, descriptive, statistical, a souvenir of the San Jose Mercury, 1896.* Reprinted, San Jose, CA: San Jose Historical Museum Association, 1986.

Simenstad, C., D. Reed, and M. Ford. 2006. "When is restoration not? Incorporating landscape-scale processes to restore self-sustaining ecosystems in coastal wetland restoration." *Ecological Engineering* 26: 27–39.

Simon, A. and M. Rinaldi. 2006. "Disturbance, stream incision, and channel evolution: The roles of excess transport capacity and boundary materials in controlling channel response." *Geomorphology* 79(3–4): 361–383.

Skinner, J. E. 1962. *An historical review of the fish and wildlife resources of the San Francisco Bay Area.* Water Projects Branch Report no. 1. Sacramento, CA: California Department of Fish and Game.

Sloan, D. 2006. *Geology of the San Francisco Bay region*. Berkeley, CA: University of California Press.

Smilie, R. S. 1975. *The Sonoma Mission San Francisco Solano de Sonoma, the founding, ruin and restoration of California's 21st mission.* Fresno, CA: Valley Publishers.

Smith and Elliott, eds. 1878. *Illustrations of Napa County, California: With historical sketch.* Reprinted,. Fresno, CA: Valley Publishers, 1974

Soderholm, E. P. 2000. *Some Napa County highlights*. Napa County Historical Society sketch no. 14. www.napahistory.org/publications.html.

Soderholm, E. P. n.d. *20,000 years of transportation to the Napa Valley*. Napa County Historical Society Sketch no. 12. www.napahistory.org/publications.html.

Sommer, T., M. L. Nobriga, W. C. Harrell, W. Batham, and W. J. Kimmerer. 2001. "Floodplain rearing of juvenile Chinook salmon: Evidence of enhanced growth and survival." *Canadian Journal of Fisheries and Aquatic Sciences* 58: 325.

Sork, V. L., F. W. Davis, P. E. Smouse, V. J. Apsit, R. J. Dyer, J. F. Fernandez, and B. Kuhn. 2002. "Pollen movement in declining populations of California Valley oak, *Quercus lobata*: Where have all the fathers gone?" *Molecular Ecology* 11(9): 1657–1668.

Southern Pacific Company. 1896. *California game "marked down"—scenic mountain woodland coverts, and tide marsh resorts for game, lakes and streams for trout, and the generous Pacific for all desirable marine contributions to sporting life.* San Francisco: Southern Pacific Company, Passenger Department.

Sowers, J. M. and S. M. Thompson. 2005. *Creek and watershed map of Milpitas and north San Jose.* Oakland, CA: Oakland Museum of California.

Spring and Lewis. 1932. *Vegetation type map, Napa quadrangle.* Courtesty of Barbara Allen-Diaz, UC Berkeley.

Squibb N. L. 1861. *Testimony in U.S. v. Salvador Vallejo*, Land case no. 39 ND [Vallejo]. United States District Court, Northern District of California. Courtesy of The Bancroft Library, UC Berkeley.

Stanford, B., R. M. Grossinger, R. A. Askevold, A. A. Whipple, E. E. Beller, M. N. Salomon, C. J. Striplen, and R. A. Liedy. Forthcoming. *East Contra Costa historical ecology study.* Oakland, CA: San Francisco Estuary Institute.

Stanly, J. A. 1883. "Some experiences in the reductions of salt marshlands." *Pacific Rural Press*. June 16.

State Census Data Center. 2010. *Census 2010: Redistricting data (public law 94-171) summary file*. California Department of Finance, Demographic Research Unit.

Stevenson, R. L. 1884. *The silverado squatters*. London: Chatto & Windus.

Stewart, O. C., H. T. Lewis, and K. Anderson. 2002. *Forgotten fires: Native Americans and the transient wilderness*. Norman, OK: University of Oklahoma Press.

Stillwater Sciences. 2006. *Napa River fisheries monitoring program final report*. Berkeley, CA: Stillwater Sciences.

Stillwater Sciences and W. E. Dietrich. 2002. *Napa River basin limiting factors analysis: Final technical report*. Berkeley, CA: Stillwater Sciences.

Stone, F. 2003. End notes and commentary by Floyd Stone. In *Napan records tales told by Vallejo descendant*. Originally published in *Napa Sunday Journal*, November 15, 1953.

Stromberg, J. C., K. J. Bagstad, J. M. Leenhouts, S. J. Lite, and E. Makings. 2005. "Effects of stream flow intermittency on riparian vegetation of a semiarid region river (San Pedro River, Arizona)." *River Research and Applications* 21(8): 925–938.

Stromberg, J. C. and D. T. Patten. 1992. "Response of *Salix lasiolepis* to augmented stream flows in the Upper Owens River." *Madroño* 39(3): 244–235.

Sullivan, C. L. 2008. *Napa wine: A history from mission days to present* (2nd edition), eds. E. Kushner, S. Ferris, and B. Imelli. South San Francisco: The Wine Appreciation Guild.

Swetnam, T. W., C. D. Allen, and J. L. Betancourt. 1999. "Applied historical ecology: Using the past to manage for the future." *Ecological Applications* 9(4): 1189–1206.

Swett, I. L., J. Aitken, and C. Harry. 1975. *The Napa Valley route—electric trains and steamers*. Glendale, CA: Ira L. Swett.

Swinchatt, J. and D. G. Howell. 2004. *The winemaker's dance: Exploring terroir in the Napa Valley*. Berkeley, CA: University of California Press.

Tappe, D. T. 1942. *The status of beavers in California*. State of California, Department of Natural Resources, Division of Fish and Game.

Taylor, B. 1862. *At home and abroad*. New York: G. P. Putnam.

Thomas, J. C. 2004. "Riparian conservation in California wine country: A comparison of the county planning approach." MA thesis, University of California Berkeley, Berkeley, CA.

Thompson, A. W. 1857. *Field notes of the final survey of the Rancho Caymus, George C. Yount, confirmee*. General Land Office, U.S. Department of the Interior, Bureau of Land Management Rectangular Survey, California, Township 6 North, Range 5 West, vol. G3. Courtesy of Bureau of Land Management, Sacramento, CA.

Thorne J., J. Kennedy, T. Keeler-Wolf, J. Quinn, J. Menke, and M. McCoy. 2004. "A new vegetation map of Napa County using the manual of California vegetation classification and its comparison to other digital vegetation maps." *Madroño* 51(4):343-363.

Tietje, W. 2011. "Urban oaks enhance wildlife diversity." In *Oaks in the Urban Landscape*, 118. Richmond, CA: University of California Agriculture and Natural Resources.

Tortorolo, M. J. 1978. "History of the city of Napa water supply." *Gleanings* 2(2): 1–12. Courtesy of Napa County Historical Society.

Tracy, C. C. 1858a. *Field notes of the final survey of Rancho Yajome, Salvador Vallejo, confirmee*. General Land Office, U.S. Department of the Interior, Bureau of Land Management Rectangular Survey, California, vol. G4. Courtesy of Bureau of Land Management, Sacramento, CA.

Tracy, C. C. 1858b. *Field notes of the survey of exterior lines of part of Township 6 and 7 N. R. 5 W. part of the south and west boundaries of T. 7 N.R. 6.W...and the offset section and transverse lines connected there with all of the Mount Diablo Meridian executed by C. C. Tracy, Deputy Surveyor under his contract dated January 25, 1858*. General Land Office, U.S. Department of the Interior, Bureau of Land Management Rectangular Survey, California, vol. R235. Courtesy of Bureau of Land Management, Sacramento, CA.

Tracy, C. C. 1858c. *Field notes of the final survey of the Rancho Carne Humana finally confirmed to the heirs courses of Edward Bale*. General Land Office, U.S. General Land Office, Department of the Interior, Bureau of Land Management Rectangular Survey, California field notes, roll 106. Courtesy of Bureau of Land Management, Sacramento, CA.

Tracy, C. C. 1858d. *Plat of the Rancho Yajome, finally confirmed to Salvador Vallejo: [Napa Co., Calif.]*. Land case no. 39 ND. United States District Court, Northern District. Courtesy of The Bancroft Library, UC Berkeley.

Tracy, C. C. 1859. *Field notes of the final survey of the Rancho de Napa, Salvador Vallejo, confirmee*. General Land Office, U.S. Department of the Interior, Bureau of Land Management Rectangular Survey, California, vol. 202. Courtesy of Bureau of Land Management, Sacramento, CA.

Trimble, S. W. 2003. "Historical hydrographic and hydrologic changes in the San Diego creek watershed, Newport Bay, California." *Journal of Historical Geography* 29(3): 422.

Triska, F. J. 1984. "Role of wood debris in modifying channel geomorphology and riparian areas of a large lowland river under pristine conditions: A historical case study." *Verhandlung Internationale Vereinigung Limnologie* 22(3): 1876–1892.

Turrill, C. B. 1876. "California notes." In *California as I saw it: First person narratives of California's early years, 1849–1900*. San Francisco: Edward Bosqui & Co. http://hdl.loc.gov/loc.gdc/calbk.185.

USACE (U.S. Army Corps of Engineers). 1927. *Calistoga quadrangle, California: Grid zone "G"* [topographic]. 1:62,500. U.S. Army Corps of Engineers tactical map. Washington, DC: Engineer Reproduction Plant, U.S. Army.

USDA (U.S. Department of Agriculture). 1942. Aerial photos of Napa County. Scale: 1:20,000. Courtesy of Napa County Resource Conservation District and Natural Resources Conservation Service.

USDA. 2005. *[Natural color aerial photos of Napa County]*. Ground resolution 1m. Washington, DC: National Agriculture Imagery Program (NAIP).

USDA. 2009. *[Natural color aerial photos of Napa County]*. Ground resolution 1m. Washington, DC: National Agriculture Imagery Program (NAIP).

USDA. 2010. *[Natural color aerial photos of Napa County]*. Ground resolution 1m. Washington, DC: National Agriculture Imagery Program (NAIP).

USDC (United States District Court). ca. 1840a. *Plan of Rancho Carne del Holmano: Napa Co., Calif.* Land case no. 47 ND, map B-104. U.S. District Court, Northern District of California. Courtesty of The Bancroft Library, UC Berkeley.

USDC. ca. 1840b. *Diseño del Rancho Yajome: Napa Co., Calif. Salvador Vallejo, Claimant*. Land case no. 39 ND, map A-85. USDC, Northern District of California. Courtesy of The Bancroft Library, UC Berkeley.

USDC (United States District Court). ca. 1840c. *Diseño del Rancho Tulucay: Napa Co., Calif.* Land case no. 45 ND, map D-95. USDC, Northern District of California. Courtesy of The Bancroft Library, UC Berkeley.

USDC. 1841. *Diseño del Rancho Napa: Napa Co., Calif., Boggs, Claimant*. Land case no. 78 ND, map D-158. USDC, Northern District of California. Courtesy of The Bancroft Library, UC Berkeley.

USGS (United States Geological Survey). 1902. *Napa Quadrangle, California* [topographic]. 1:25,000. Washington, DC: USGS.

USGS. 1913. *Topographic instructions of the United States Geological Survey.* Washington, DC: Government Printing Office.

USGS. 1943. *Calistoga quadrangle, California, 15-minute series.* [topographic]. 1:62,500. Washington, D.C.: USGS.

USGS. 1980a. *Mare Island Quadrangle, California: 7.5-minute series* [topographic]. 1:24,000. Washington, DC: USGS.

USGS. 1980b. *Napa Quadrangle, California: 7.5-minute series* [topographic]. 1:24,000. Washington, DC: USGS.

USGS. 1981. *Cuttings Wharf Quadrangle, California: 7.5-minute series.* [topographic]. 1:24,000. Washington, D.C.: USGS.

USSG (United States Surveyor General). 1869. *Plat of the part of the Rancho de Napa, finally confirmed to John Truebody as located by the U.S Surveyor General.* Courtesy of Napa County Surveyor's Office.

Vallejo, J. L. 1885. *J. L. Vallejo to Napa City Water Co. A true copy of an original recorded at request of Geo. E. Goodman Aug. 6th 1885, at 3 mins past 11 am. Henry Brown, Co. Recorder.* Courtesy of City of Napa.

Vallejo, S. 1861. *Testimony in U.S. v. George C. Yount.* Land case no. 32 ND. United States District Court, Northern District of California. Courtesy of The Bancroft Library, UC Berkeley.

van der Velde, G., R. S. E. W. Leuven, A. M. J. Ragas, and A. J. M. Smits. 2006. "Living rivers: Trends and challenges in science and management." *Hydrobiologia* 565: 359–367.

Vines, B. 1861. *Testimony in U.S. v. George C. Yount.* Land case no. 32 ND. United States District Court, Northern District of California. Courtesy of The Bancroft Library, UC Berkeley.

Wallace, W. 1901. *History of Napa County 1901.* Oakland, CA: Enquirer Print.

Walter, R. C. and D. J. Merritts. 2008. "Natural streams and the legacy of water-powered mills." *Science* 319: 299–304.

Waples, R. S., T. Beechie, and G. R. Pess. 2009. "Evolutionary history, habitat disturbance regimes, and anthropogenic changes: What do these mean for resilience of Pacific salmon populations?" *Ecology and Society* 14(1): 3.

Warner J. C. 2000. Barotropic and Baroclinic Convergence Zones in Tidal Channels. PhD thesis, University of California, Davis.

Weaver, C. E. 1949. *Geology and mineral deposits of an area north of San Francisco Bay, California.* Bulletin 149. San Francisco: State of California Department of Natural Resources.

Weber, L. 1998. *Old Napa Valley: The history to 1900.* St. Helena, CA: Wine Ventures Publications.

Weber, L. 2001. *Roots of the present: Napa Valley 1900–1950.* St. Helena, CA: Wine Ventures Publications.

Whipple, A. A., R. M. Grossinger, and F. W. Davis. 2011. "Shifting baselines in a California oak savanna: Nineteenth century data to inform restoration scenarios." *Restoration Ecology* 19(101): 88–101.

White, M. A. and D. J. Mladenoff. 1994. "Old growth forest landscape transitions from pre-European settlement to present." *Landscape Ecology* 9: 191–205.

White, M.D. and K. A. Greer. 2006. "The effects of watershed urbanization on the stream hydrology and riparian vegetation of Los Penasquitos Creek, California." *Landscape and Urban Planning* 74(2): 125–138.

White, R. 1995. *The organic machine: The remaking of the Columbia River*. New York: Hill and Wang.

Whitney, G. G. 1994. *From coastal wilderness to fruited plain: A history of environmental change in temperate North America from 1500 to the present*. Cambridge: Cambridge University Press.

Whitthorne, R. C. 1969. *Mosquito control in the Napa Valley*. Napa, CA: Napa County Historical Society.

Wichels, J. ca. 1976. *Centennial anniversary of Rutherford 1876–1976. A sketch from the archives of the Napa County Historical Society,* series 1, no. 3. Napa, CA: Napa County Historical Society.

Wichels, J. n.d. *Unabridged story of Yountville*. Unpublished manuscript.

Wilson, E. O. 1984. *Biophilia*. Cambridge, MA: Harvard University Press.

Winfrey, G. 1953. "Napan records tales told by Vallejo descendant." *Napa Sunday Journal*, November 15.

Wohl, E. 2004. *Disconnected rivers*. New Haven, CT: Yale University Press.

Worster, D. 1985. *Rivers of empire: Water, aridity, and the growth of the American West*. New York: Oxford University Press.

Xiao, Q., E. G. McPherson, S. L. Ustin, M. E. Grismer, and J. R. Simpson. 2000. "Winter rainfall interception by two mature open-grown trees in Davis, California." *Hydrological Sciences* 14: 763–784.

Yount, G. 1831. *Writings by George Yount, Napa Valley, 1831*.

INDEX

Numbers following "n" indicate note number.

acorn woodpecker, 26, 29 (fig.), 186n59
Adams, I. C., 85, 178
Adobe, Juarez's, 165, 166 (fig.)
agriculture
    development in Napa Valley and, 144–45
    dry farming, 15, 50, 55 (figs.), 160, 187n15
    groundwater level and, 17, 54–55, 186nn10, 12
    as habitat type in 2010, 146–47 (map), 148 (table)
    history of, 13, 14–15, 18–19, 182n28
    in reclaimed tidal marshes, 129 (fig.), 134, 135, 136, 195n27
    in wet meadows, 72, 73 (maps)
    See also orchards; ranching; vineyards; wheat cultivation
alders, 111, 193n86
    red alder, 110
alkali meadows
    in early 1800s, 22–23 (map)
    in early 1800s vs. 2010, 78, 148 (table)
    as habitat type, 21 (table)
    overview of, 68, 76–77, 77 (figs.), 188n16
alkali milkvetch, 77
alluvial fans
    agriculture on, 26, 30, 163
    braided channels on, 62, 63
    floodplain defined by, 92 (map), 92 (fig.), 106, 163
    freshwater marshes formed by, 70
    impact on Napa River form, 82, 116–17 (figs.), 175
    landscape tours of, 167, 173
    overview of, 9 (fig.), 10
    spreading streams on, 52
Altimira, José, 18, 34, 182n25
American Canyon
    landscape tour near, 155, 158 (map), 159–61
    population, 11, 13, 17
    tidal marshland restoration, 124
arroyo willow, 110
ash (trees), 111, 193n86
Atwater, B. F., 193n4
Avery, Benjamin Parke, 36 (fig.), 111

backwater. *See* sloughs
Baker, Milo, 74
Bale Creek, 64 (fig.), 175
Bale Mill, 64 (fig.), 175, 177
Bale Slough, 70, 79 (fig.), 170 (map), 173, 173 (fig.), 188n3
Barnett, Elias, 88
Bartlett, John Russell
    on agriculture in Napa Valley, 14
    on boats sailing to Napa, 122, 159
    as data source, 18
    on lack of undergrowth, 34
    on Napa River, 82, 85, 110
    on oaks, 26, 32, 36, 43 (fig.), 167
    wildflowers observed by, 38, 159
Bartlett, W. P., 32
basket sedge, 110, 188n9
bass, striped, 114
bat, Pacific pallid, 26, 186n59
Bay Area Aquatic Resources Inventory (BAARI), 78, 106
bay laurel, California, 60–61 (fig.), 193n86
bears, 69
    grizzly bear, 26
beavers
    in Napa River, 108 (figs.), 108, 109 (fig.), 192nn66, 70
    Napa River dry season flow and, 191n52
    salmon restoration and, 108, 192n68
    in sloughs, 89
Bell Creek, 50
Bickford, E. L., 88 (fig.)
big-leaf maple, 110
bird's beak, soft, 123
blackberries, 69, 188n3
blackbird, tricolored, 70 (fig.), 159
black oak, California, 36, 185nn32, 37
black-tailed deer, 50
Borelli-Zavala, Josanna, 167
Bothe–Napa Valley State Park, 175, 177
bottlebrush, 44 (fig.)
Bowles, Samuel, 34
bracted popcorn flower, 74
braided channels
    in early 1800s, 22–23 (map)
    as habitat type, 21 (table)
    landscape tour view of, 174
    overview of, 62 (fig.), 62–63, 63 (figs.)
Brewer, William, 26, 122, 128
Brewster, Elise, 167
brickell bush, California, 110
bridges
    1st Street Bridge, 113 (fig.), 165
    Garnet Creek Bridge, 48–49 (fig.)
    Pope Street Bridge, 100–101 (figs.), 110
    Pratt Avenue Bridge, 97 (fig.)
    3rd Street Bridge, 165
    Trancas Street Bridge, 167
    Zinfandel Lane Bridge, 96 (figs.), 174
Briggs, L. Vernon, 112
brook trout, eastern, 114
Bryan, E. N., 187n15
buckeye, California, 177, 177 (fig.), 193n86
Bucknall, Mary E., 86, 191n59
Bull Island, 136, 138–39 (maps), 139, 161, 195n35
burning. *See* pyro-management

burr oak. *See* valley oaks
Byrns, Robert, 134–35

calicoflowers, 74
California bay laurel, 60–61 (fig.), 193n86
California black oak, 36, 185nn32, 37
California brickell bush, 110
California buckeye, 177, 177 (fig.), 193n86
California clapper rail, 123
California Department of Fish and Game, 124
California lilac, 193n86
California poppy, 38, 38–39 (fig.), 185n43
California rose, 111
California State Coastal Conservancy, 17, 124
California sycamore, 110
Calistoga
  alkali meadows, 76, 76–77 (figs.)
  elevation, 5
  freshwater marshes, 70, 71 (fig.)
  hot springs, 9, 24 (fig.)
  landscape tour near, 175, 176 (map), 177–78
  oaks in/near, 3 (fig.), 24–25 (fig.), 29 (fig.), 36, 178 (fig.)
  population, 17
Calistoga popcorn flower, 74
Canijolmanok, Chief, 174
Canijolmano tribe, 11, 145
Carneros, landscape tour near, 155, 158 (map), 161–62
Carneros Creek, 60, 64, 158 (map)
Carpenter, E. J.
  on agriculture, 182n28, 187n15
  on alluvial fans, 10
  on natural levee of Napa River, 186n6
  recommended increasing drainage, 54, 56
  on saltgrass, 188n16
  soil surveys, 19, 72, 188n7
  on spreading streams, 52
  on tidal marshland vegetation, 128, 129
  on types of trees in savannas, 185n37
Carvelli, Walter, 128, 129
cattle. *See* ranching
Caymus land grant (Rancho), 12, 52, 85, 102, 189n16
Caymus tribe, 11, 14, 145, 169, 172
channel incision. *See* incision, channel
Charles Krug winery, 40, 174 (fig.), 174
chat, yellow-breasted, 58, 106 (fig.), 168–69 (fig.)
Chinook salmon, 83, 83 (fig.), 114, 115 (table)
chub, thicktail, 68, 114
Chum salmon, 115 (table)
cinnamon teal, 70 (fig.), 74
clapper rail, California, 123
Clean Water Act, 17, 84
climate, 5, 8, 68, 181n10
climate change, 21, 149, 151, 182n41
clovers, 38
  owl's clover, 38, 38–39 (fig.)
  sack clover, 77
  showy Indian clover, 72, 159

Clyman, James, 70, 128
coast live oak, 36, 185nn33–34
Coho salmon, 115 (table)
collinsias, 38
Cone, Mary, 32, 36
Conn Creek
  braided channel of, 9, 62, 63
  landscape tour of, 170 (map), 171, 172–73
  Napa River and, 64, 88, 102
  reservoir on, 13, 50, 98, 103, 190–91n38
Coombs, Nathan, 52, 88, 105
Cosby, Stanley W.
  on agriculture, 182n28, 187n15
  on alluvial fans, 10
  on natural levee of Napa River, 186n6
  recommended increased drainage, 54, 56
  on saltgrass, 188n16
  soil surveys, 19, 72, 188n7
  on spreading streams, 52
  on tidal marshland vegetation, 128, 129
  on types of trees in savannas, 185n37
cottonwoods, 111, 193nn83, 85
  Fremont cottonwood, 110
cream cups, 38
creeks, 48–65
  altered drainage network of, 56, 57 (map)
  with braided channels, 21 (table), 22–23 (map), 62 (fig.), 62–63, 63 (figs.)
  contemporary views of, 50–51 (figs.)
  early maps of, 51 (map), 52 (map)
  high water table and, 54–55, 186n6
  incision/erosion on, 60–61 (figs.), 60–61, 187n24
  overview of, 50–51
  riparian vegetation along, 58, 59 (figs.), 187n19
  spreading flow of, 52, 53 (maps), 186n6
  summer water, 48–49 (fig.), 64 (fig.), 64, 64–65 (map), 187n34, 188n40
  See also specific creeks
Cronise, T. F., 185n36
cuckoo, yellow-billed, 58, 104, 106 (fig.), 191n56
Curtis, Edward S., 35 (fig.)
Cyrus, John, 86

dams
  beaver, 108, 108 (fig.)
  on creeks, 50, 63, 98, 103
  diversion, for dry season water, 102, 167, 191n47
  natural knoll dams, 71 (fig.), 175
  possible removal of, 51
Darms, H. A., 131
Davis, Cornelius Eiling, 136
deep bay/channel
  in early 1800s, 22–23 (map)
  in early 1800s vs. 2010, 146–47 (map), 148 (table)
  as habitat type, 21 (table)
deer, 34, 104
  black-tailed deer, 50

Delta tule pea, 123, 129
Dewoody, T. J., 53 (fig.), 85, 104–5, 105 (fig.), 194n15
Digger pine. *See* gray pine
diked marsh, as habitat in 2010, 146–47 (map), 148 (table)
Douglas fir, 175
Douglas' meadowfoam, 74 (fig.)
drainage
    altered network of, 56, 57 (map)
    cut off by Napa River natural levee, 54, 186n6
    geology's impact on, 8, 9 (fig.), 9, 10
    See also groundwater level
Dry Creek
    contemporary view of, 50 (fig.)
    landscape tour of, 164 (map), 167, 169
    oaks on fan of, 43 (figs.), 167
    spreading flow, 53 (map), 90 (fig.), 163
    summer flow, 64
dry farming, 15, 50, 55 (figs.), 160, 187n15
ducks, 135, 140, 161, 195n32
    mallards, 74
    ruddy duck, 70 (fig.)
    wood duck, 26, 83, 88 (fig.), 89, 104, 189n18
Duhig, Stewart, 140, 161
dwarf calicoflower, 74

eastern brook trout, 114
Eastwood, Alice, 74
elk, 34, 86, 104, 123
    tule elk, 26
Elliott, Wallace W.
    atlas as data source, 18
    on dry farming, 187n15
    on oaks, 26, 32, 183n3
    on riparian vegetation, 193n86
    on sources of summer water, 187n34
    on St. Helena, 40, 174
erosion. *See* incision, channel
eucalyptus, 44 (fig.), 129 (fig.), 147 (map), 161

Fagan Marsh, 135, 138–39 (maps), 139, 148
fans. *See* alluvial fans
fire. *See* pyro-management
firs, 36
    Douglas fir, 175
fish, native
    complex channel morphology supporting, 100–101 (figs.)
    in creeks, 50, 51
    importance of flooding to, 92
    in Napa River, 83, 83 (fig.), 114 (fig.), 114, 115 (table)
    See also specific fish
Fisher, C. K., 96
flooding
    in 1955, 19, 92, 92 (map), 92–93 (fig.), 190n22
    importance of, to native fish, 92
    overview of, 83, 84 (fig.), 189nn2, 4
    See also Napa River Flood Protection Project
Foster, D. R., 181n2

Fremont cottonwood, 110
freshwater marshes
    building roads through, 69
    contemporary, 69 (fig.)
    in early 1800s, 22–23 (map)
    in early 1800s vs. 2010, 78, 79 (map), 146–47 (map), 148 (table)
    as habitat type, 21 (table)
    Napa River spread into, 86, 86 (map), 87 (map)
    origin of, 78
    overview of, 68, 70, 70 (fig.), 71 (figs.)
Friends of the Napa River, 4, 13, 17
frogs, 169
    red-legged frog, 70 (fig.)
    tree frog, 50

Garnet Creek, 48–49 (fig.)
geography, 5, 8, 144–45
geology, 6–7 (maps), 8–10, 9 (fig.), 10 (fig.), 181n12
Gilbert, G. K., 194nn4–5
Glass Mountain, 9
glides, 98
goby, tidewater, 114
goldfields, 32 (fig.), 74
grains. *See* wheat cultivation
grapes, wild, 83, 110, 111 (fig.), 193n86
    See also vineyards
grassland/wildflower field
    in early 1800s, 22–23 (map)
    in early 1800s vs. 2010, 146–47 (map), 148 (table)
    as habitat type, 21 (table)
gravel bars
    along Napa River, 98, 99 (fig.), 100–101 (figs.), 110 (fig.), 133, 190nn36, 38
    on braided channels, 62, 63 (fig.)
    on landscape tour, 165, 165 (fig.)
gravel mining, 62 (fig.), 63, 98, 187n24, 190nn36, 38
Gray, N., 85, 104, 105 (fig.)
gray pine, 36, 185n37
Green Island Road, 159
Grigsby, Mary, 190–91n38
Grinnell, J., 103, 108, 189n18, 195n32
grizzly bear, 26
groundwater level
    agriculture and, 17, 54–55, 186nn10, 12, 187n15
    artificial drainage altering, 56, 57 (map)
    Napa River's natural levee and, 186n6
    valley oaks and, 54
    See also drainage
gumplant, Pacific, 122

habitats
    climate change and, 21, 149, 151, 182n41
    in early 1800s, 21 (table), 22, 22–23 (map)
    in early 1800s vs. 2010, 146–47 (map), 148 (table)
    See also specific habitats
Harley, J. B., 181n2
harvest mouse, salt marsh, 123, 140

Hathaway, V., 195n27
hazelnut, 193n86
Hedel, C. W., 193n4
Hennessey, Lake, 13, 15
Higuera, Nicolas, 85
historical ecology
    benefits of using, 145, 148–49
    composite historical map, 20 (table), 20–21, 21 (table), 22–23 (map)
    data sources, 17–19
    overview of, 2, 4, 181nn1–2, 5
Hittell, John S., 40, 195n22
Holmes, L. C., 72, 188n7
Horseshoe Bend, 161

Illustrations of Napa County, California (Smith and Elliott), 18, 26
incision, channel
    efforts to reverse, 118, 119 (fig.)
    Napa River, 96 (figs.), 96–97, 97 (fig.), 190n33
    overview of, 60–61, 60–61 (figs.), 187n24
Indian clover, showy, 72, 159
indigenous people. *See* native Californians
interior live oak, 185n33
irrigation
    history of, 14–15, 17
    lack of, 181n10
    reservoirs constructed for, 50
    unnecessary for some crops, 55 (figs.), 187n15
    using tidal waters, 128

Jepson, Willis
    conifers observed by, 36
    as data source, 19
    on indigenous pyro-management, 184n28
    on oaks, 26, 45 (fig.), 46
    on riparian vegetation, 110 (fig.), 110, 111
    wildflowers observed/collected by, 38, 72, 74, 185n44, 188n9

Kashiwagi, J. H., 19, 72, 188n7
Kerr, David
    led survey team (1858), 18
    spreading streams on maps of, 53 (map)
    tidal marshlands map, 85, 126, 126–27 (map), 136–38 (maps), 194n6
    on trees in Napa Valley, 113, 184n17
Kimball Creek, 50–51 (fig.), 98
Knights Valley, 32 (fig.)
knolls, 9, 70, 71 (figs.), 82, 171
Kunkel, F., 187n15, 191n52

lagoon, as habitat type in 2010, 146–47 (map), 148 (table)
Lambert, G., 19, 72, 188n7
landscape tours, 152–78
    lower valley, 163–69, 164 (map)
    mid-valley, 170 (map), 170–74
    south of Napa, 155–62, 156 (map), 158 (map)
    upper valley, 175–78, 176 (map)
landscaping, native, 47 (fig.)

Land Trust of Napa County, 13, 136
Lapham, Macy, 14
larks, 193n2
laurel, California bay, 60–61 (fig.), 193n86
Leach, Frank, 58, 165
Learned, Babe, 40, 183n3
Leidy, Robert, 114
Leopold, L. B., 187n19
levees
    around tidal marshlands, 124, 125, 134–35, 136, 139 (fig.)
    constructed to reduce flooding, 14, 83, 130
    natural, of Napa River, 9 (fig.), 52, 116–17 (fig.), 186n6
Light Detection and Ranging (LIDAR), 116
lilac, California, 193n86
lilaeopsis, Mason's, 123, 129
live oaks
    coast live oak, 36, 185nn33–34
    interior live oak, 185n33
    landscape tours of, 169, 172, 178 (fig.)
    prevalence of, 36 (fig.), 36, 185n37
    in riparian corridors, 111, 193n86
    in tidal marshlands, 122
Living River principles, 17, 182n39
lupine, 38 (fig.), 38–39 (fig.)
Lyman, William, 177
Lyman Memorial Buckeye Grove, 177 (fig.), 177

madronas, 185n36
main stem/channel
    defined, 189n14
    dredging, 130, 131 (fig.), 132–33 (maps), 195n20
    length of, 190n20
    straightness/sinuosity of, 86, 94, 94 (table), 95 (maps), 190nn27, 29
mallards, 74
Man and Nature (Marsh), 2, 181n5
manzanitas, 169, 193n86
maple, big-leaf, 110
maroonspot calicoflower, 74
Marple, William Lewis, 120, 120–21 (fig.), 160
Marsh, George Perkins, 2
marshes. *See* freshwater marshes; tidal marshes
Mason, Herbert, 74
Mason's lilaeopsis, 123, 129
Mayacma tribe, 11
McFarling, John, 86
meadowfoam, Douglas', 74 (fig.)
Mendell, G. H., 130
Menefee, Campbell Augustus, 34, 50, 111, 134, 185n36, 193n86
milkvetch, alkali, 77
Mill Creek, 33 (fig.), 176 (map), 177 (fig.), 177
Milliken, R. A., 181n18
mission/rancho period, 11, 12, 14, 18, 27, 182n25, 183n2
Money, Lydia M., 136
monkeyflower, yellow, 72
mosquitoes, 135, 195n33
Motzkin, G., 181n2

mountains, 9, 10 (fig.), 181n12
mouse, salt marsh harvest, 123, 140
Muybridge, Eadweard, 26, 178

Napa (city)
    dry season water source, 102, 191n47
    elevation, 5
    images over time, 1 (figs.), 15 (fig.)
    landscape tour in/near, 163, 164 (map), 165 (fig.), 165–67, 166 (fig.)
    population, 12–13
    problematic location, 195n22
    slough, 89 (map)
Napa County, population growth, 12–13
Napa County Agricultural Preserve, 13, 15
Napa County Flood Control and Water Conservation District, 17, 124
Napa County Resource Conservation District, 17
Napa Creek
    landscape tour of, 164 (map), 165
    Napa River vs., 85
    riparian vegetation, 58, 110 (fig.), 110, 187n20
    summer flow, 64–65 (map)
Napa Golf Course, 162
Napa Green certification, 17
Napa River, 80–119
    altered network draining into, 56, 56 (fig.), 57 (map)
    beavers in, 108 (figs.), 108, 192nn66, 70
    changed character, 1 (figs.), 82–84, 189nn7–8
    changed depth, 56 (fig.), 96 (figs.), 96–97, 97 (fig.), 190n33
    dredging, 130 (fig.), 130, 131 (fig.), 132–33 (maps), 195n20
    dry season flow, 102 (fig.), 102–3, 103 (table), 191nn43, 47, 50, 52
    fish in, 83, 83 (fig.), 114 (fig.), 114, 115 (table)
    freshwater flows from Delta into, 129, 194n19
    gravel bars along, 98, 99 (fig.), 100–101 (figs.), 110 (fig.), 190nn36, 38
    islands in, 88, 189n16
    Napa Valley topography and, 116, 116–17 (figs.)
    natural levee of, 9 (fig.), 52, 116–17 (fig.), 186n6
    response to improving health of, 118 (figs.), 118–19 (figs.), 119 (fig.)
    as river vs. creek, 85, 85 (maps), 85
    sinuosity of, 86, 94 (table), 94, 95 (maps), 190nn27, 29
    snags and logjams in, 108–9, 192n74
    spread into wetlands, 86 (map), 86, 87 (map)
    tidal river corridor, 112, 112–13 (figs.)
    See also riparian forest; flooding; sloughs
Napa River Ecological Reserve, 83 (fig.), 87 (fig.), 107 (map), 107 (fig.), 168–69 (figs.), 169
Napa River Flood Protection Project
    guiding principles, 17, 182n39
    history of, 83
    planted riparian vegetation, 112 (fig.), 162
    restored channel complexity, 89 (map), 130–31
    restored tidal marshlands, 136, 136 (map)
Napa River Rutherford Reach Restoration Project, 17, 144, 189n8
Napa River Trail, 84, 162
Napa Sustainable Winegrowing Group, 144
Napa tribe, 11

Napa Valley
    changing landscape, 3 (figs.), 144–45
    in early 1800s, 20 (table), 20–21, 21 (table), 22–23 (map)
    future of, 150–51 (figs.), 151
    influence of topography, 116, 116–17 (figs.)
    land-use timeline, 11–13, 182n24
    map of, 6–7 (map)
    Santa Clara Valley vs., 15, 16 (figs.), 16
    sources of data on, 17–19
    study area, 7 (map)
Napa Valley Historical Ecology Project, 4
Napa Valley Vintners, 17
Napa Watershed Information Center and Conservancy, 13, 17
narrowleaf willow, 110
native Californians
    in Napa Valley, 11, 12, 35 (fig.), 134, 181n18
    place names, 14
    pyro-management by, 11, 14, 34, 182n25, 184nn25, 27–28
native landscaping, 47 (fig.)
Nelson, J. W., 72, 188n7
newts, 50
Nolte, George, 19
nuthatch, white-breasted, 26
nutsedge, 110

Oak Knoll
    agricultural crops, 3 (fig.)
    landscape tour near, 163, 164 (map), 167
    Napa River summer flow, 102
    oaks in area, 36, 42, 42–45 (figs.), 186n56
    vernal pool complexes, 74, 75 (figs.)
Oak Knoll Farm, 18, 43 (fig.)
oaks, 3 (fig.)
    black oak. *See* California black oak
    burr oak. *See* valley oaks
    California black oak, 36, 185nn32, 37
    coast live oak, 36, 185nn33–34
    interior live oak, 185n33
    mapping historical distribution, 30 (figs.), 30–31, 31 (map), 183nn10, 12–13, 184n14
    in modern landscape, 45 (figs.), 46, 47 (figs.), 186n59
    in riparian corridors, 111, 193n83
    scattered throughout savannas, 32 (fig.), 32–33, 33 (figs.), 184n17
    trajectories of, near Oak Knoll, 42, 42–45 (figs.), 186n56
    weeping oak. *See* valley oaks
    white oak. *See* valley oaks
    willow-oak. *See* valley oaks
    See also live oaks; valley oaks
oak savannas, 24–47
    cattle ranching under, 27, 183n2
    in early 1800s, 22–23 (map)
    in early 1800s vs. 2010, 146–47 (map), 148 (table)
    ecological role, 26–27, 28–29 (fig.), 183n2
    as habitat type, 21 (table)
    historical accounts, 24–25 (fig.), 26, 27, 30, 32, 183nn2–3
    historical photos, 24–25 (fig.), 29 (fig.)

oak savannas *(continued)*
    mapping, 30 (figs.), 30–31, 31 (map), 183nn10, 12–13, 184n14
    pyro-management of, 11, 14, 34, 184nn25, 27–28
    scattering of oaks in, 32 (fig.), 32–33, 33 (figs.), 184n17
    settlements in, 40, 40–41 (fig.), 185n52
    trajectories of, 42, 42–45 (figs.), 46 (fig.), 186n56
    types of trees, 36, 40, 185nn31–34, 36–37
    wheat grown under, 27, 28–29 (fig.), 183n5
    wildflowers in, 32 (fig.), 38, 38–39 (figs.)
Oakville
    landscape tour near, 170 (map), 171, 172
    slough near, 90 (fig.)
Old Faithful Geyser, 69 (fig.), 177, 177 (fig.)
oleander, 44 (fig.)
open water, artificial, as habitat type in 2010, 146–47 (map), 148 (table)
orchards
    dry farming, 50, 55 (figs.), 160, 187n15
    high water table problem for, 54, 186n10
    history of, 3 (fig.), 13, 14
    oaks and, 27, 42, 47 (fig.)
    wildflowers in, 38–39 (fig.)
Oregon ash, 110
Osborne, Joseph, 43 (fig.)
overflow. *See* sloughs
owl's clover, 38, 38–39 (fig.)
"the Oxbow," 3 (fig.), 94, 113 (fig.), 166
Oxbow bypass, 89 (map), 165

Pacific gumplant, 122
Pacific pallid bat, 26, 186n59
Palisades rock formation, 9
Parry's tarplant, 38 (fig.)
Patchett, John, 13
Patwin, 11
pea, Delta tule, 123, 129
Peattie, D. C., 183n3
perch, Sacramento, 114
perennial freshwater pond
    in early 1800s, 22–23 (map)
    as habitat type, 21 (table)
photography, 19
    See also Turrill and Miller photographs
phylloxera, 13, 14
pickleweed, 122, 126 (fig.)
pines
    Digger pine. *See* gray pine
    gray pine, 36, 185n37
    ponderosa pine, 36, 175
Pometta, Inez, 38 (fig.)
ponderosa pine, 36, 175
pond turtle, western, 159
popcorn flowers, 74, 74 (fig.), 185n44
    bracted popcorn flower, 74
    Calistoga popcorn flower, 74
    rusty popcorn flower, 74
    slender popcorn flower, 74
poppy, California, 38, 38–39 (fig.), 185n43

population growth, 11–13, 15, 17, 182n32
pronghorn, 26, 34
pyro-management, 11, 14, 34, 182nn25, 184nn25, 27–28

quail, 193n2

rabbit, tule, 195n35
Rackham, O., 181n2
rails, 193n2
    California clapper rail, 123
rainbow trout, 114, 115 (table)
rainfall, 5, 181n10
ranching
    history of, 12–13, 14
    riparian vegetation and, 50, 129 (fig.)
    in shade of oaks, 27, 183n2
Raven, Peter, 74
Rector Creek, 50, 50–51 (fig.), 98, 170 (map), 172
red alder, 110
red-legged frog, 70 (fig.)
red willow, 110
redwoods, 111
Regional Water Quality Control Board, 17
Revere, J. W., 110
riparian forest
    composition of, 110–11, 111 (fig.), 193nn81, 83–86
    in early 1800s, 22–23 (map)
    as habitat type, 21 (table)
    images of, 80–81 (fig.), 82 (fig.), 110 (fig.)
    incision undercutting, 60–61 (figs.), 60–61
    loss and recovery of, 58, 59 (figs.), 187n19
    narrowing of, 106, 107 (fig.), 107 (map), 192n63
    in past vs. present, 104 (fig.), 104–5, 105 (fig.), 146–47 (map), 148 (table)
    in tidal corridor, 112, 112–13 (figs.)
    wildlife requirements for, 104, 106 (fig.), 191n56
Ritchey Creek, 58, 59 (figs.), 64 (fig.), 82, 103, 176 (map)
Rodgers, Augustus F., 109
rose, California, 111
ruddy duck, 70 (fig.)
run-pools, 98
Russians, 12, 14
rusty popcorn flower, 74
Rutherford, 69, 170 (map), 171, 172–73
Rutherford Reach Restoration Project, 17, 144, 189n8

sack clover, 72
Sacramento perch, 114
Sacramento splittail, 123, 129
salmon, 89
    beavers' importance to, 108, 192nn68
    Chinook salmon, 83, 83 (fig.), 114, 115 (table)
    Chum salmon, 115 (table)
    Coho salmon, 115 (table)
    floodplain and wetlands' importance to, 92
    in tidal marshes, 123
saltgrass, 76 (fig.), 128, 188n16

salt marsh harvest mouse, 123, 140
salt ponds
    as habitat type in 2010, 146–47 (map), 148 (table)
    landscape tours of, 159
    marshland restoration and, 134 (fig.), 136, 137 (fig.), 138–39 (maps), 139
San Francisco Bay, tidal marshlands, 123, 193–94n4
San Francisco Estuary Institute (SFEI), 4
Santa Clara Valley
    Napa Valley vs., 15, 16 (figs.), 16
    oaks in, 30, 183n9, 184n17
Santa Ynez Valley, 184n17
Sauer, Carl, 2, 181n5
Scouler's willow, 110
scrub jay, 45 (fig.), 148
secondary channels. *See* sloughs
sedges, 128
    basket sedge, 87 (fig.), 110, 188n9
    nutsedge, 110
shallow bay/channel
    in early 1800s, 22–23 (map)
    in early 1800s vs. 2010, 146–47 (map), 148 (table)
    as habitat type, 21 (table)
Sharpsteen Museum, 178
shining willow, 110
showy Indian clover, 72, 159
side channels. *See* sloughs
sinuosity, of Napa River, 86, 94 (table), 94, 95 (maps), 190nn27, 29
Skinner, J. E., 192n66
skunk, striped, 50
Slaughterhouse Point, 157 (fig.), 157
slender popcorn flower, 74
sloughs
    animals in, 88 (fig.), 89, 134
    landscape tours of, 157, 165
    other terms for, 88
    overview of, 88 (map), 88–89, 89 (fig.), 89 (map), 90–91 (figs.), 189n15, 190n20
    tidal sloughs, 122, 126–27 (map), 139
    See also Bale Slough
smartweed, 110
Smith, Clarence L.
    atlas as data source, 18
    on dry farming, 187n15
    on oaks, 26, 32, 183n3
    on riparian vegetation, 193n86
    on sources of summer water, 187n34
    on St. Helena, 40, 174
snipe, 193n2
soft bird's beak, 123
soil
    oak distribution and, 30, 183n10
    USDA surveys of, 72
    in wet meadows, 72, 188n7
spearscale, San Joaquin, 77
splittail, Sacramento, 123, 129
spreading streams, 52, 53 (maps), 186n6

Spring Creek, 63 (fig.)
Squibb, Nathaniel L., 86 (fig.), 86, 90 (fig.), 189n10
Stanly, John A., 128, 194nn7, 15, 19
Stanly Lane, 161
star tulip, large-flowered, 72
steelhead
    migrating in creeks, 50, 51
    in Napa River pools, 8383 (fig.)
    prevalence of, 114, 114 (fig.), 115 (table), 148
    in tidal marshlands, 123
    where to see, 174
Stevenson, Robert Louis, 26, 32, 140, 159, 163
St. Helena
    landscape tour near, 170 (map), 171, 173–74
    population, 11, 13, 17
    swimming hole, 102 (fig.)
    valley oaks, 40
    York Creek dam, 51
St. Helena, Mount, 10 (fig.)
striped bass, 114
Sulphur Creek
    alluvial fan, 30 (fig.), 92 (fig.), 106
    braided channel of, 9, 62, 62 (fig.), 63, 63 (figs.)
    landscape tour of, 171, 174, 176 (map)
    riparian vegetation along, 58, 58 (fig.), 148
Suscol, 160 (fig.), 160–61, 195n22
Suscol Creek, 30, 52 (map), 158 (map), 160
sycamores, 111, 185n36, 193nn81, 83–84
    California sycamore, 110

Tappe, D. T., 192n68
tarplant, Parry's, 38 (fig.)
Tavernier, Jules, 66–67 (fig.)
Taylor, Bayard, 128
teal, cinnamon, 70 (fig.), 74
terroir, 144, 145
thicktail chub, 68, 114
Thompson, A. W., 40, 85, 102, 104, 110, 111, 193n83
tidal flats
    in early 1800s, 22–23 (map)
    in early 1800s vs. 2010, 146–47 (map), 148 (table)
    as habitat type, 21 (table)
tidal marshes, 120–41
    dominated by tules, 128–29, 129 (fig.), 194nn10–11, 15, 19
    dredging river channel in, 130 (fig.), 130, 131 (fig.), 132 (map), 132–33 (map)
    in early 1800s, 22–23 (map)
    in early 1800s vs. 2010, 146–47 (map), 148 (table)
    as habitat type, 21 (table)
    Kerr's map of, 126, 126–27 (map), 194n6
    landscape tours of, 157, 157 (fig.), 159, 162
    landward edge, 140 (fig.), 140, 141 (fig.)
    overview of, 122–24, 123 (fig.)
    plants and animals, 122–23, 135, 193n2, 195n33
    before reclamation, 120–21 (fig.), 124–25 (map)
    reclamation of, 129 (fig.), 134–35, 136, 194n5, 195nn27, 32–33
    restoration of, 130–31, 134 (fig.), 135, 136, 136–39 (maps), 139, 195n35

tidal marshes (continued)
    of San Francisco Bay, 123, 193–94n4
    tides in, 126, 194nn5, 7
    trajectory of, 123–24, 135 (map), 135, 195n34
tidal marshlands. See tidal marshes
tides
    lands reopened to, 135, 136, 136 (fig.), 139, 148 (table)
    in tidal marshlands, 122, 126, 129, 162, 194n7
tidewater goby, 114
tidy-tips, 32 (fig.), 38 (fig.)
timelines
    data sources, 17–19
    land use, 11–13, 182n24
toothed calicoflower, 74
toyon, 193n86
Tracy, C. C., 69, 85, 105, 106
Trancas Crossing Park, 84, 167
Trancas Street Bridge, 167
tree frog, 50
Trefethen Family Vineyards, 27 (fig.)
tricolored blackbird, 70 (fig.), 159
trout
    eastern brook trout, 114
    rainbow trout, 114, 115 (table)
True, Elijah, 160
tule elk, 26
tule pea, Delta, 123, 129
tule rabbit, 195n35
tules, 120 (fig.), 123, 128–29, 129 (fig.), 169, 194nn10–11, 15, 19
Tulucay Creek, 50–51 (fig.), 53 (map), 158 (map), 164 (map)
Turrill, Charles, 19
Turrill and Miller photographs
    alkali meadows in Calistoga, 76–77 (fig.)
    as data source, 19
    Napa Creek near St. Helena, 110 (fig.)
    oaks, 28–29 (figs.), 36–37 (figs.), 40–41 (fig.)
    upper Napa River, 82
turtle, western pond, 159

Upson, J. E., 187n15, 191n52
urban/barren/vacant habitat, as habitat type in 2010, 146–47 (map), 148 (table)
urban development, 15 (fig.), 15, 17, 182n32
U.S. Army Corps of Engineers, 83, 130 (fig.), 130–31, 131 (fig.), 132–33 (maps)
U.S. Fish and Wildlife Service, 124

Valencia, Manuel, 80–81 (fig.)
Vallejo (city)
    landscape tour near, 155, 156 (map), 157, 157 (fig.), 159
    population, 155, 182n32
Vallejo, Ignacio, 34, 69, 83, 123, 182n25
Vallejo, Salvador, 88, 174
valley freshwater marsh. See freshwater marshes
valley oaks
    alternative names, 36, 46, 185n31
    canopy of, 28–29 (figs.)

groundwater level and, 54
historical images, 33 (fig.), 35 (fig.), 36–37 (fig.)
landscape tours of, 161, 162, 162 (fig.), 167, 172–73 (fig.), 173, 174, 174 (fig.), 178 (fig.)
mapping historical distribution of, 31 (map), 183nn12–13
in modern landscape, 27 (figs.), 44–45 (figs.), 46, 47 (figs.)
removal of, 27, 183nn3, 5
in riparian corridor, 61 (fig.), 110, 111
in Santa Clara Valley, 183n9
in savannas, 24–25 (fig.), 36, 36–37 (fig.), 40, 185n37
settlements under, 40, 40–41 (fig.), 185n52
trajectories of, 42, 42–45 (figs.), 186n56
wheat grown under, 27, 28–29 (fig.), 183n5
as witness trees, 36
valley oak savanna. See oak savannas
valley wetlands. See wetlands
vernal pool complexes
    in early 1800s, 22–23 (map)
    in early 1800s vs. 2010, 78, 148 (table)
    as habitat type, 21 (table)
    overview of, 68, 74, 75 (fig.)
    wildflowers in, 74, 74 (figs.)
vernal pool fairy shrimp, 159
Vines, Bartlett, 64, 68, 70, 102
vineyards
    climate and, 5, 8, 181n10
    early, 3 (fig.), 13, 14, 19
    geology and, 9
    irrigation of, 17, 187n15
    in mid-Napa Valley, 171
    Napa Green certification, 17
    phylloxera problem, 13, 14
    present-day, 13, 17
    terroir and, 144–45
Vischer, Edward, 18, 33 (fig.), 35 (fig.)
volcanic activity, 9

Wallace, W. F., 64
Wappo
    in Napa Valley, 11, 35 (fig.)
    pyro-management by, 34, 182n25, 184n27
Warner, J. C., 194n19
watercress, 188n9
water table. See groundwater level
Watkins, Carleton, 24 (fig.), 26, 178
Weaver, C. E., 190n36
weeping oak. See valley oaks
West, W. P., 134
western pond turtle, 159
wetlands, 66–79
    in early 1800s vs. 2010, 148 (table)
    historical maps, 68 (maps)
    Napa River spread into, 86, 86 (map), 87 (map)
    overview of, 66–67 (fig.), 68–69
    past vs. present, 78, 79 (map), 189n20
    See also alkali meadows; freshwater marshes; tidal marshes; vernal pool complexes; wet meadows

wet meadows
    in early 1800s, 22–23 (map)
    in early 1800s vs. 2010, 78, 146–47 (map), 148 (table)
    as habitat type, 21 (table)
    landscape tours of, 159
    origin of, 78
    overview of, 68, 72, 73 (maps), 188nn7, 9
wheat cultivation
    history of, 12–13, 14, 182n28, 183n5
    landscape tour of areas, 173
    under oaks, 27, 28–29 (fig.), 183n5
Whipple, A. A., 184n17
white-breasted nuthatch, 26
white oak. *See* valley oaks
White Slough marshlands, 157
white sturgeon, 123
wildflowers
    landscape tours of, 159
    in meadows surrounding wetlands, 66–67 (fig.)
    in oak savannas, 32 (fig.), 38, 38–39 (figs.)
    in vernal pool complexes, 74, 74 (figs.)
    See also specific wildflowers
wild grapes, 83, 110, 111 (fig.), 193n86
Williams, Thomas H. Jr., 133
Williams, Virgil, 32 (fig.)
willow-oak. *See* valley oaks
"The Willows," 91 (fig.), 105, 106, 169, 191n59
willows, 110, 111, 112 (fig.), 120–21 (fig.), 122, 193n86
    arroyo willow, 110
    narrowleaf willow, 110
    red willow, 110
    Scouler's willow, 110
    shining willow, 110
    yellow willow, 110
wineries. *See* vineyards
witness trees, 18, 36, 184n30
wood duck, 26, 83, 88 (fig.), 89, 104, 189n18
woodpecker, acorn, 26, 29 (fig.), 186n59
Work, John, 108

yellow-billed cuckoo, 58, 104, 106 (fig.), 191n56
yellow-breasted chat, 58, 106 (fig.), 168–69 (fig.)
yellow monkeyflower, 72
yellow willow, 110
York Creek, 51
Yount, George
    on California poppies, 38, 185n43
    and Caymus tribe, 169, 172
    granddaughter, 86, 191n59
    grave, 169
    Rancho Caymus land grant, 12
Yountville
    landscape tours near, 163, 164 (map), 169, 170 (map), 171, 172
    population, 11, 13, 17
    sloughs, 88, 90–91 (figs.)

Zinfandel Lane, 173–74

wet meadows
    in early 1800s, 22–23 (map)
    in early 1800s vs. 2010, 78, 146–47 (map), 148 (table)
    as habitat type, 21 (table)
    landscape tours of, 159
    origin of, 78
    overview of, 68, 72, 73 (maps), 188nn7, 9
wheat cultivation
    history of, 12–13, 14, 182n28, 183n5
    landscape tour of areas, 173
    under oaks, 27, 28–29 (fig.), 183n5
Whipple, A. A., 184n17
white-breasted nuthatch, 26
white oak. *See* valley oaks
White Slough marshlands, 157
white sturgeon, 123
wildflowers
    landscape tours of, 159
    in meadows surrounding wetlands, 66–67 (fig.)
    in oak savannas, 32 (fig.), 38, 38–39 (figs.)
    in vernal pool complexes, 74, 74 (figs.)
    See also specific wildflowers
wild grapes, 83, 110, 111 (fig.), 193n86
Williams, Thomas H. Jr., 133
Williams, Virgil, 32 (fig.)
willow-oak. *See* valley oaks
"The Willows," 91 (fig.), 105, 106, 169, 191n59
willows, 110, 111, 112 (fig.), 120–21 (fig.), 122, 193n86
    arroyo willow, 110
    narrowleaf willow, 110
    red willow, 110
    Scouler's willow, 110
    shining willow, 110
    yellow willow, 110
wineries. *See* vineyards
witness trees, 18, 36, 184n30
wood duck, 26, 83, 88 (fig.), 89, 104, 189n18
woodpecker, acorn, 26, 29 (fig.), 186n59
Work, John, 108

yellow-billed cuckoo, 58, 104, 106 (fig.), 191n56
yellow-breasted chat, 58, 106 (fig.), 168–69 (fig.)
yellow monkeyflower, 72
yellow willow, 110
York Creek, 51
Yount, George
    on California poppies, 38, 185n43
    and Caymus tribe, 169, 172
    granddaughter, 86, 191n59
    grave, 169
    Rancho Caymus land grant, 12
Yountville
    landscape tours near, 163, 164 (map), 169, 170 (map), 171, 172
    population, 11, 13, 17
    sloughs, 88, 90–91 (figs.)

Zinfandel Lane, 173–74